The Electric Wilderness

Facsimile Edition
2011

Books by Andrew A. Marino

Electromagnetism & Life [Facsimile Edition]

Modern Bioelectricity

Going Somewhere

Further publications available at
andrewamarino.com

THE ELECTRIC WILDERNESS

Andrew Marino and Joel Ray

With a Preface by Robert Becker

Cassandra Publishing
Belcher, LA

Printed in the USA

ISBN 978-0-9818549-2-2

Publisher's Cataloging-in-Publication Data:

To Linda and Marilyn

PREFACE

The development of electromagnetic energy for power and communications is widely regarded as a boon to mankind and even as "the manifestation of civilization at its finest."* Although one cannot dispute that mankind's mastery of this silent force is most responsible for our present global technology, disquieting questions have arisen over its present safety and the wisdom of continued unlimited expansion in its use. The electromagnetic force is, like the nuclear force, one of the four basic physical forces of the universe. While the ego of some men exults in "mastering" giant forces, have we once again acted like children playing with fire? Thus far we have been able to keep the nuclear genie for the most part in the bottle, but the electromagnetic genie is out of the bottle and all around us. There is no place left on earth that is free of manmade, abnormal electromagnetic radiation. Have we already done perhaps irretrievable harm to the biota, or is there time to step back, look at what we have done, and plan for a safe, logical, and environmentally sound use of this silent force?

When I set in motion the affair described in this book with my letter to Joseph Swidler, then head of the New York Public Service Commission, I naïvely believed that the questions I raised over the possible hazards of powerline radiation would be settled by the logical, dispassionate methods of science. As *The Electric Wilderness* shows, that was most certainly not the case. Our arguments, based on what we considered clear scientific data, were met with a ferocious onslaught—not only against our data but our reputations and even our livelihoods. Legal, political, and bureaucratic maneuvering extended from the local level to the highest levels of the Federal government. Why? Was it simply that we were calling into question a favorite technological application and with it the egos of the engineers, or that our actions might diminish the profits of the industries concerned?

In my opinion we faced a concerted and coordinated effort to suppress the truth, which emanated from the military establishment and was simply aided and abetted by the greed of the utilities and the tarnished testimony of scientists for hire. Today the military is planning and constructing the largest expansion of electromagnetic emitting facilities in history, with the aim of fighting and winning the next World War. In the process the very population, culture, and civilization they are sworn to protect may be placed at risk. Although someone has apparently weighed the relative risks and decided that this was the path to follow, the decision has been made not only without input from the citizens, but with every effort to conceal the risk from them. It seems the same mind-set that led to disaster in Vietnam is still operating—that "in order to save the village from Communism, we have to destroy it"—only now the villagers are not our "enemies" but our own citizens.

---

* Review by Samuel C. Florman of *Networks of Power* by Thomas P. Hughes (Johns Hopkins, 1984), in *The Sciences*, New York Academy of Sciences, 1984.

The decisive information that would help citizens contest this expansion will not be available; as this is being written, the decision has been made to terminate all Federal funding for electromagnetic hazards research. Thus those of us still committed to this fight are effectively deprived of the conclusive proof of hazard which has just begun to emerge. If a serious hazard does exist, no one is to know about it, least of all the citizens of the village. If opposition to a facility develops, no matter how small or obscure, the full weight and power of the establishment will be brought to bear to insure that it is constructed, and that all questions of health hazards are submerged in a sea of doubletalk and outright deception.

The citizens of this country are poorly served by the present system. They lack funds, battalions of lawyers, paid scientific experts, and organization. Their one source of power, concerted action at the ballot box, is continuously subverted by misinformation and distortion. Meanwhile the issue is obfuscated, the public is convinced that no hazard exists, and the scientists who insist on open and public debate are unmercifully attacked.

This book is important not only in the issue it raises of the probable hazard of electromagnetic technology, but in the hazards it reveals that are the result of *raising the issue*. It spotlights the present defects in our political system that leave the citizen defenseless against the powerful manipulations of the government. The truly important questions of our time are those relating to technology and its uses and abuses in relation to human beings. The public has a right to the relevant scientific information, and cannot be denied the right to have a voice in decisions that affect their health, safety, and quality of life. However, though scientists can provide the information, any scheme that calls for "experts" to make the decisions of relative risk tends to have little value. The only valid and ethical risk analysis must be made by those who are at risk. That requires that citizens have access to the truth. *The Electric Wilderness* is the truth.

Robert O. Becker, M.D.

Lowville, N.Y., 1985

AUTHOR'S NOTE

The collaboration with Andy Marino which led to this book began, for me, well before we actually met—reading his remarkable testimony in the New York hearing, then later watching twelve friends get arrested for resisting the State's "right of eminent domain" at the Barses' farm in Fort Covington, N.Y. By instinct and profession I was deeply impatient at the violations of language and common sense with which the more rapacious technologies (modern prisons, nuclear plants, atomic weapons) were defended. But what seems today a foregone conclusion regarding my involvement as a writer in the powerline issue was certified only when, in March 1977, I saw Andy in action on the witness stand. Here, I recognized, was a man who not only understood the real issue, but had the scientific wherewithal and especially the will to make that understanding stick. It was, and is, a rare combination of qualities.

It took a while after we decided to write the book to realize that the story had to be told in the first person. Although we planned and wrote it together, the experience it described was Andy's; the form of the book had to reflect that.

My understanding of science and policy issues—and of the human issues behind them—has been greatly broadened by work on this book. I hope the reader, too, will learn something of value from this experience as our society tries to come to grips with the destructive effects of its expertise.

Joel Ray

Ithaca, N.Y.

TABLE OF CONTENTS

DEFINITION

NIEMR *(nee mer), n.* acronym for nonionizing electromagnetic radiation; a form of electrical energy emitted by transmission lines, radio and TV towers, microwave ovens, radar, electric blankets, CB radios, etc. *Synonyms:* microwaves, electromagnetic waves, electric waves, electric signals, electric rays, electric fields, magnetic fields, electromagnetic fields.

Note: Although there are many variations in types of NIEMRs (in frequency, strength, waveform, etc.), the generic term is used in this book. For further clarification, the reader is encouraged to consult the sources listed in the Selected Reading List beginning on p. 111.

PART I

CHAPTER 1

The Meeting

When my boss, Dr. Robert Becker, returned to the lab from the Washington meeting, a new and troubling concern was taking shape in his mind.

It was December 1973. Three months before, at a conference in New York City, Becker had been approached by U.S. Navy Commander Paul Tyler with a request to serve on a panel of experts to evaluate some experiments the Navy had funded. It all had to do with an antenna system the Navy was planning to build in northern Wisconsin, a bizarre project involving grids of buried wires that would extend over thousands of square miles of land there. The Navy had been studying ways of communicating with submerged submarines for years; the project, called Sanguine, was believed to be the answer. (It was to be later renamed SEAFARER, and still later ELF, an acronym for Extremely Low Frequency.) But because of the large size of the antenna system, and fears that the NIEMR it would emit might have environmental and health impacts, Congress had ordered the Navy to undertake a series of studies to see whether there would be any problems. All the studies involved exposing living systems to Sanguine-type NIEMR. The program, Tyler had told Becker, was now at its midpoint: initial studies had been completed, and what the Navy needed was expert opinion on the meaning of the results.

Tyler must have realized that of all the scientists attending the conference—sponsored by the New York Academy of Sciences—Becker was among the best qualified to answer the Navy's questions. And that was in fact true. For fifteen years Becker had been conducting research into the relation between electrical and magnetic forces and living things, and the three-day Academy conference, entitled Electrically Mediated Growth Mechanisms in Living Systems, was in large measure a result of his work. It was the first time that American scientists had been gathered under such prestigious auspices to explore the subject of bioelectricity; the tributes to Becker were frequent. After he had delivered the keynote paper of the second day, one scientist had acknowledged that "we have learned so much from you over the years, in fact, almost everything that we know about this field"; another had called Becker's paper "one of the most significant in the history of human biology."

1

Becker's 1973 paper had been the grand summation of his work up to that time. His interest was in growth and regeneration, and in 1958 he had set out to determine how these key biological phenomena occurred. In a way, his interest was an embodiment of the question that had puzzled scientists from the beginning: What is the difference between "living" and "dead"?

It had not taken Becker very long to uncover some important clues. Though he had no lab at the time—he was Chief of Orthopedic Surgery and mainly concerned with clinical matters—U.S. Veterans Administration (VA) hospitals had been gearing up their research programs in response to the Soviet launch of Sputnik in 1957, and he had asked for some funds from the VA Board for a small experiment. He had been reading Luigi Galvani, the 18th century Italian anatomist, and had noted that in an anonymously published experiment Galvani had observed that injury was accompanied by an electrical current. Becker had also read Albert Szent-Gyorgyi's 1941 lecture in which the Nobel prizewinner had hypothesized that the cell might have electrical properties crucial to certain biological functions. Then, after reading a startling Russian paper that reported the regeneration of tomato plants by electricity, Becker requested $1000 from the VA Board to do an experiment with electricity and animals.

Soon things began to move fast. In the early 1960s Becker obtained the first evidence for his hypothesis that healing and growth were controlled by tiny self-generated electrical signals which mobilized and directed cellular activity. By 1964, around the time I went to work for him, he had postulated the existence of a previously undescribed electrical communications system within living things. These were revolutionary ideas in biology, and they generated a lot of controversy.

Becker's main interest in these ideas was medical. He reasoned that if one could find out how the body heals itself, one should be able to gain control over that process and reproduce it in cases where healing broke down for some reason. But in addition to the medical implications, there were other questions as well, questions that had to do with the origins of life and with the relationship between living things and the natural electromagnetic environment.

If there were these weak neural currents inside living things, did they provide a way for animals and humans to interact with the changing magnetic field of the earth? Becker's research was many-sided. In 1963 he found that the number of admissions to psychiatric hospitals seemed to vary according to changes in the earth's magnetic field. And in 1967 he reported that weak magnetic fields applied to the heads of humans altered their reaction time.

Shortly after that, Becker did a very interesting experiment in which he applied currents to frog blood cells in glass chambers and found that they reverted to a more primitive cell type. Then he was able to regenerate partially the foreleg of a rat with electrical current. Regeneration was a power that amphibians possessed—salamanders could grow back whole limbs— but not mammals. It was one of the most controversial experiments Becker ever did; there are still scientists today who do not believe it.

By 1973 he had begun to use tiny electrical currents applied through implanted electrodes to cure human bones that would not heal normally. Thus it was just at the time of the New York Academy conference that the therapeutic implications of Becker's work were beginning to take shape. His research at that point was on bone, but perhaps the techniques of electrotherapy might be eventually applied to soft tissue as well, such as heart muscle. The horizon, in 1973, seemed unlimited.

However, Becker worried that things might get out of hand. Electrotherapy had been used in the nineteenth century, but because of the fraudulent practices of certain charlatans, among other things, belief in the therapy had been undermined. It would be terrible if history were to repeat itself. Physicians might use electrotherapy irresponsibly, before it was fully understood, and people might get hurt. Becker reasoned that if electricity could cause benign growth, it might also cause malignant growth. Several times at the New York conference, Becker warned the other scientists that caution was the watchword. A few bad mistakes with patients and electrotherapy might once again be discredited. He had not spent fifteen years of difficult and often heavily criticized research to see it all go down the drain because some doctor, overanxious for money or fame, induced cancer in a patient he was trying to heal. (For a full account of Becker's research, see his recent book, co-authored with Gary Selden, *The Body Electric: Electromagnetism and the Foundation of Life*, William Morrow, 1985.)

*

Because of the worry about the inadvertent side effects of electrotherapy, and because of some experiments I had just finished that showed adverse effects in mice and rats from exposure to certain kinds of NIEMR, Becker was supersensitive to the potential implications of the Navy antenna. When he returned to the lab that Monday in December he told me what the Navy studies had found.

There had been seven scientists on the panel. After describing the Sanguine antenna, Tyler told them that the Navy needed to know where to go from there with further studies. He spent most of the first morning reading the results of the more than thirty studies; nearly two-thirds of them had found biological effects from exposure to the NIEMR, in a variety of species including slime mold, rats, birds, and human beings. Becker had felt he would probably be in the minority on this panel, because in 1973 very few thought that energy of such low frequency and strength could cause biological effects. But what happened surprised him. Before long it became clear that *all* the panel members were thinking the antenna was a potential hazard to human health.

On the second day, the panel began to draw up a lengthy list of recommendations. It was clear there had to be further study, especially on humans. Becker interpreted some of the results as due to a lowering of the body's normal resistance, a kind of NIEMR-induced stress, similar to what had happened to the mice and rats in our own study.

In the middle of the deliberations someone pointed out that the Sanguine NIEMR was similar in nature to that produced by high-voltage powerlines, and that in the largest lines, those of 765,000 volts, the strength of the NIEMR might be as much as *a million times stronger*. That realization, said Becker, had really thrown the group into a quandary. They had been

3

asked there to review studies on the antenna. What could they do about powerlines, which might represent an even greater potential hazard, involving many more people? The discussion became heated, but eventually the scientists agreed they had to recommend some action by the government. Their concluding recommendation was that the Navy should inform a special committee advisory to the President that many Americans might be "at risk" from powerline NIEMR.

But Becker's concern didn't end there. For when he returned home that weekend he read in the newspaper that the Power Authority of the State of New York (PASNY—a kind of state TVA) planned to build a 765,000-volt line from the Canadian border to Utica, and that the route would take it not far from where some of his friends lived. He had noticed all the powerlines and radio and TV and microwave towers on his way to the airport, he said—all emitting forms of NIEMR—but he remembered telling the panel that there were no really big powerlines in New York. The newspaper notice drove home the point, once and for all, that people *were* "at risk."

I didn't pay as much heed to these ramifications as to the fact that the Navy studies seemed to confirm what we were finding; I guess my main reaction at the time was elation. Here we were doing experiments with very little money, and the Navy was coming up with similar results, after the expenditure of millions of dollars. I'm sure that aspect of it pleased Becker too, and he was no doubt thinking that we might get some funding from the Navy. Money was always a problem in the VA lab.

Anyway, I settled down to design another series of experiments, and the NIEMR hazard disappeared from my mind, until six months later.

CHAPTER 2

Simpson Arrives

The new experiments we decided to do were in connection not with pow-erlines but with Becker's bone-healing work. There were a couple of rea-sons for them. Because the beneficial effects of the electricity on Becker's patients might conceivably be due not to the NIEMR itself but, say, to the metal in the electrodes, we had to test this possibility by using NIEMR with-out wires in experiments on animals. Instead of implanting electrodes, we would expose animals to NIEMR that involved no contact whatsoever with tissue. The other reason was that it might be possible to develop a therapy that did not use electrodes, a "noninvasive" therapy. Before Becker could try it with humans, we had to check it out on animals.

So I designed an experiment that would expose rats to NIEMR for thirty days. We would compare these animals with control rats that lived in ex-actly the same circumstances without exposure, and see what the differ-ences were. It was a low-budget experiment, and I built the apparatus in my basement at home, using leftover scrap wood and metal—shelving, panel-ing, and trim—from when I had remodeled my living room. Esthetically the result wasn't particularly appealing, but functionally it was perfect. It fit smoothly with the regular procedures for animal care at our facility, and I used it in experiments with over 500 animals for more than thirty months and had no significant problems. When the time came to build the power supply to provide the NIEMR, cost was again the overriding consideration, so I used the cheapest source of power available—electricity from the wall outlet, with suitable provisions to protect against the danger of shock.

Soon after the study began, I saw changes in the exposed rats that were not occurring in the controls. The exposed animals seemed to gain less weight and drink less water, and they had altered levels of blood proteins and enzymes. Since the results were clearly adverse, they tended to fore-close the use of certain kinds of NIEMR in our clinical work, and that was pretty much what I saw as their value.

But when Becker looked, he saw something more. He saw powerlines. The NIEMR I was using was, after all, from electricity derived from a wall outlet, and that was at a frequency of 60 Hertz (cycles per second)—the fre-quency of the entire electrical distribution system in North America. Becker asked me what the results meant with regard to 765,000-volt lines. I said I hadn't the foggiest idea. Did I know what the strength was in relation to that of such lines? I did not; I guessed one would have to be very close to the wires—maybe within a foot or two—to experience the same strength I had used in the study, but I really didn't know for sure. "Well," said Becker, "find out." And as he left he said, "Do the experiment again."

5

I repeated the experiment twice; the results were the same.

Then, in July, Becker told me he had just had a call from a lawyer with the New York Public Service Commission (PSC), who wanted to come and see him about a powerline hearing. That's when I learned that, immediately after he had returned from Washington and the Navy meeting, Becker had written a letter to several state regulatory agencies, to tell them that the powerline planned by PASNY might be a health hazard. He alerted them to the Navy studies and urged them to explore the issue before they allowed the line to be built. I asked whether he'd had any response, and he said an engineer from Niagara-Mohawk (NiMo), the local utility, had called and said rather rudely that he thought Becker was raising a bogus issue. The agencies had not responded until now. It seemed that *another* 765,000-volt line was being planned for the Rochester area, by NiMo and Rochester Gas and Electric (RG&E), and that the NIEMR issue had been raised in that hearing by a citizens' group.

Becker asked me if I would sit in on the meeting with the lawyer, whose name was Bob Simpson.

As I say, powerlines were the furthest thing from my mind. And in a way it's odd that I didn't feel more strongly about the connections Becker was making, because when it came to other forms of pollution I had strong feelings. In fact, three years earlier, partly as a result of a public hearing on a quarry operation that was screwing up my neighborhood, I had decided to go to law school so I could learn how to argue environmental cases. But it was the well-known kinds of pollution I was thinking of then—air and water pollution, food additives, pesticides, and so on. I loved my research work with Becker, but I also had ambitions about going into court some day and arguing the case that would establish the constitutional right to a clean environment. Also, I was married, and beginning a family, and I was concerned about what kind of world my kids were going to have to deal with. When Simpson called to see Becker, I had been out of law school for a month, and I was wondering how I could begin using my legal training—but not in relation to powerlines.

Simpson arrived on 15 July 1974, in the morning. He was younger than I, and had been out of law school only a few years. His only job had been on the PSC staff. He told us a little about the public hearing in Rochester. The experts testifying for the power companies had said there would be no hazard from the NIEMR, but the citizens' group had refused to accept their assurances and had found evidence enough so that the issue ought to be seriously addressed. Simpson showed us the testimony of the main company expert. I thought it was not only inaccurate but downright ignorant. For example, the expert had said that cows under a 765,000-volt line were contented because he had seen their tails wagging.

Simpson asked, was there really a potential health risk? "No question about it," said Becker. He detailed for Simpson the adverse effects that the Sanguine program had turned up, and said that the powerline NIEMR would be far stronger. He also told him about the experiments I was doing on the rats, and that they had been twice repeated with the same results.

Becker is a proud, direct man who likes to be recognized for what he knows and what he can do. As with the request from Paul Tyler to come to Washington, it was flattering to be asked by a government representative to help him do his job—especially when his job was to help protect the public interest. Sensing this trait, Simpson asked Becker whether he would be willing to testify about the possible health risk in the hearing. Becker was a teacher—the best I had ever encountered—and this was the role in which he felt most comfortable; it was that side of him to which Simpson was appealing.

Becker wanted to know the details. Simpson answered in a way that understated the work that would be involved, and I could see that Becker was rising toward a decision to testify. Though the citizens' group had uncovered some preliminary information, they scarcely had the expertise to testify as Becker could, and I think Becker understood that. Simpson said Becker's testimony would consist of a written report to be transmitted to the power companies, and later perhaps a day of cross-examination in Albany by the power company lawyers. After that, PSC would decide whether the 765,000-volt line could be built, and if so, how it would have to be regulated.

Simpson seemed wary of me. He hadn't known that I would be sitting in on the meeting, and when he found out that I had just finished law school, he seemed surprised; he was probably juxtaposing "biophysicist" and "lawyer" and wondering exactly what I was and where I fit in here. His invitation to testify was extended only to Becker.

But Becker suggested that both of us should testify. After all, I had actually done the rat studies, and I had in-depth training in the mathematics and physics of electricity (which is why he had hired me in the first place). Also he may have been thinking that my legal knowledge would be helpful, though I doubt that was foremost in his mind. (He had arranged things so that I could go to law school, but he really had little use for lawyers.)

The legal aspect of the hearing certainly interested me. In fact as I listened to Simpson and considered the power company testimony that had already been given in the case, I had an insight into the way the legal system worked that shook me a bit because it seemed so obvious and yet I hadn't really considered it. It had to do with evidence. The point was that judges made their decisions on the basis of evidence presented to them. If some piece of information was "in evidence" it could be used in deciding, but if it wasn't "in evidence" it couldn't. Now the scientific merit of the companies' testimony was abysmal. But the judge didn't know that, and the citizens' group had no real scientific expertise and neither did Simpson. The net result would be the elevation of ignorance and grossly self-serving statements to the level of fact, so as to support an inevitable decision that the line would be safe. What I realized that day as I listened to Simpson was, simply, that if you control the evidence, you control the outcome.

I wondered what it would be like working with Simpson. He troubled me a little. First, he as much as admitted that he'd been assigned to the case because he was the low man on the PSC totem pole. His real interest was utility rate regulation. I couldn't imagine *anything* more dull or boring than that, and it didn't make me especially confident in him. How could he be expected to handle a case involving arcane scientific concepts?

The other question was why he was asking Becker to testify. The Sanguine committee report had been classified For Official Use Only, so Becker didn't feel he could properly use it in a hearing. Why not ask Tyler to testify, or at least the investigators who had done the actual experiments? As I saw it, all *we* could present were articles in the scientific literature—if there were any—and our own experiments, which were preliminary. That didn't stack up well against all the expensive Navy work. I asked Simpson about it. He said bluntly that he had asked as many people as he could identify, and they had all refused; he didn't really know why.

In the end, Becker established that we would both testify. Simpson said we would be paid for our time, but Becker instinctively realized that would be a bad idea, because it would restrict our independence. He told Simpson no, we would do it for free. That turned out to be a very wise decision.

The only perk Becker requested was that on the day we were to be cross-examined the hearing should be held in Syracuse. He had patients to consider. Simpson said he saw no difficulty in arranging it. And that was that.

*

During the next few months Becker and I worked on our reports for Simpson to send the power companies. I found eight studies published in the open scientific literature that described biological effects in animals or humans from exposure to NIEMR similar to that of a powerline. I described these studies, and also my own rat experiments, and with the help of a former classmate who had become a physics professor, I calculated the strength of the NIEMR from the proposed line at various measured distances from the wires. As I had expected, it became progressively weaker with increasing distance from the line. What I didn't expect was the great distance to which perceptible radiation would extend—several thousand feet. This calculation also allowed me to relate my rat study to the powerline. I found that the strength I had used there was comparable to the strength at chest height directly underneath the wires.

Becker's testimony began with a brief recounting of his research over the past fifteen years, to show the firm scientific foundation on which his opinions rested. His medical conclusions were that the existing data showed NIEMR to be a biological stressor, and that as a physician he would have to assume the effects would be harmful.

It would be unethical, Becker wrote, to expose people to levels of NIEMR greater than ambient levels without their permission. Such exposure would be tantamount to human experimentation without informed consent. Before he, as a doctor, could expose human beings to such fields, he had to comply with every provision of the Human Experimentation Regulations, which included obtaining their informed consent and telling them that they could terminate their participation at any time. It seemed to Becker absurd that power companies could expose people without their consent, in the pursuit of profit.

We finished the reports in October and Simpson mailed them to the power companies. It was the first time ever in the U.S. that scientists—moreover, scientists with all the right degrees, experience, and requisite professional affiliations—had presented hard evidence in an official forum that a health risk could result from exposure to powerlines.

The reaction of the power companies to our testimony was immediate and telling. They requested that the Rochester hearing be suspended so they could produce new witnesses. And soon after, things began to develop in such a way that we knew we had opened a real can of worms.

CHAPTER 3

Escalation

Becker's 1973 letter to the agencies, which ultimately led to Simpson's visit, had been in response not to the RG&E/NiMo powerline (he hadn't even known about it at the time), but to the one planned by the Power Authority (PASNY). At the time our testimonies were mailed, PASNY's hearing before the PSC was well under way and proceeding smoothly. PASNY was saying that there was no danger from the NIEMR. There were groups in northern New York also opposed to the PASNY line, but they had not been able to convince the PSC hearing officer—as the Rochester group had—that NIEMR might be a problem. Indeed, PASNY seemed so certain their line would be approved that during the hearing they had begun to negotiate purchases of materials and contracts for the purchase of power from Quebec. Looking at the transcripts from that hearing, I got the feeling that the hearing examiner in charge, a man named Thomas Matias, was simply going through the motions. I assumed that he had read Becker's letter and was simply ignoring it. For a judge, Matias was disturbingly casual. When a farmer voiced fears that birds flying near the line might be electrocuted, Matias replied, "Well, then, we won't have to shoot them."

Our testimonies for the RG&E/NiMo hearing wound up changing PASNY's plans as well. As a result things began to get much more complicated for me and Becker.

Once we had raised the NIEMR issue, the bureaucrats in the state agencies realized that it would make no sense for one hearing examiner to conclude that there was a health risk because of our testimony there, and for another to conclude that there was no health risk from an identical line because we did not participate in *that* hearing. The feeling began to develop that there had to be some sort of proceeding that would settle the matter once and for all—a kind of generic hearing. But whereas RG&E and NiMo were content with a delay that would enable them to prepare a new and better defense, PASNY was not. They were a state authority, for one thing, and for years had been used to getting their way. Moreover, the line from the Quebec border, which would tie into an enormous future supply of electricity from the huge James Bay hydro project, had already been granted certain federal exemptions on the basis of its importance to national security. Also, PASNY's line was planned for operation much sooner than the Rochester line. So PASNY began to apply pressure on PSC for an expedited decision.

Then came word that the agencies were beginning to squabble among themselves about how to deal with the unfamiliar health issue we had raised. Besides PSC and PASNY, The Department of Environmental Conservation (DEC), the Attorney General, and the Department of Agriculture and Markets were jockeying for position; many fundamental questions emerged.

Not only was the issue new, but the type of hearing envisioned was unprecedented as well. Which agency should in fact be responsible for overseeing the state's inquiry into the health and safety of high-voltage powerlines? Which issues should be raised or considered in the hearing? Other issues had emerged in the Rochester hearing: electrochemical pollution, TV and radio interference, electric shock, effects on cardiac pacemakers, noise. Should all these factors be considered, in addition to the question of human exposure to the NIEMR? What procedures should be established? What was the best way, for instance, of making the other five companies in New York parties to the proceeding so that when and if they proposed to build 765,000-volt lines, the entire hearing wouldn't have to be repeated?

In the end, a broad agreement was reached. The RG&E/NiMo application would be joined with that of PASNY, and all companies in the state would be given an opportunity voluntarily to participate, which meant that if they did, the results would be binding on them. What the establishment of a new hearing meant for Becker and me was brand new testimony. Because the inquiry had expanded so dramatically, because so much more was riding on what we said, these new documents would have to be even stronger than before. At least, though, we would have a full year to develop them, because PSC decided to adopt a two-step procedure. In Step One, all the physical characteristics of the lines would be established. This part would involve mainly engineering testimony to determine the amount of noise the line would make (especially in foul weather), the amount of ozone it would produce, the type and extent of radio and TV interference, the levels of current induced in large metal objects under the line, and, most important, the strength of the NIEMR at various distances from the wires. This step of the hearing would take about one year. Then Step Two would address the crucial question: what dangers would these physical realities pose for humans? PSC ruled that no further lay testimony would be admitted, only that of experts. The judges in charge of the new hearing would be Lawrence Gollomp, from the Rochester hearing, and Thomas Matias. Later, Gollomp would be removed and Matias would become the chief judge.

The expanded hearing procedure made no one happy, except of course the citizens who would have to live near the lines. The whole purpose of the Rochester citizens' group in pressing the NIEMR issue had been to uncover scientists who could testify for them, and that had been accomplished the day Simpson came to see us. PASNY officials were very upset; with their own hearing close to finished, the new hearing was a big step backward for them. RG&E, one of the power companies most hostile toward state bureaucracy and regulation—on general principles, it seemed—now found its fate directly tied to that of a state agency, one with resources that dwarfed its own. The power companies that had no plans for 765,000-volt lines saw no reason at all for a hearing; after all, there were thousands of miles of such lines in operation and no evidence that anyone had ever been hurt by them. DEC wanted to *force* every power company to participate in the hearing, not just invite them in as PSC wanted to do. The Health Department was ambivalent throughout. It felt that our testimony involved public health matters and should be presented under its aegis. However, that would be difficult to arrange in the context of a state licensing procedure that had been established previously by the legislature, and in any case the Health Department admitted that it had no expertise to make judgments on the NIEMR issue. So in the end it acceded to the PSC plan, with the recommendation that Becker and I be sponsored by the PSC staff. That meant the Attorney General could not participate because he had no witnesses.

For Becker and me, the expanded procedure was just plain scary. We had started out to testify in one local hearing, involving one line and two upstate companies, and now our potential adversaries were the entire New York State utility system, including a powerful and virtually unregulated state authority. At one point we learned that even one out-of-state company was going to join in. We had submitted our October 1974 testimony under what looked to be dispassionate conditions. It had seemed reasonable to me that developments outside the orbit of the electric power industry could have gone unnoticed by the industry, and I saw our role as providing information from just such an unanticipated source to those in the industry with the wherewithal to make necessary changes to protect the public. I actually expected to be treated with respect, if not gratitude, for performing this public service.

But that's not the way things were working out. In bits and pieces over the next few months, we began to learn that we were to be the objects of a witch-hunt. Rumors began to circulate that the industry was going to "get Becker and Marino," and Marino in particular was to be the target. Though our experiments had been jointly planned, I was the one who actually performed them, who searched the literature for other reports of comparable studies which showed effects, and who contacted other investigators working on NIEMR bioeffects. I was the dangerous one.

But it wasn't just the power industry reaction that bothered me. Even within PSC, the agency we were to testify for, the wheeling and dealing caused serious problems. One that outraged me involved Simpson's boss, Arthur Rheingold, who was the PSC Assistant General Counsel. Simpson called me one day, dejection in his voice, to say that Rheingold was going to take over the case. I did not trust Rheingold. I thought he was attracted by the idea of handling two potential superstars, and by the likelihood of taking expense-paid trips around the country to drum up other witnesses; I did not want him representing me in the courtroom. I told Simpson to tell Rheingold that if Simpson didn't stay, he could kiss me goodbye as a witness. Luckily, the threat worked, and Simpson stayed.

I was pretty unsure where I stood when I did that, and it was a risk. But something told me I had to stand firm. Now I know it was one of the most important things I ever did. By early 1975 Simpson and I were developing a good rapport, and he kept me abreast of the politics of the hearing. Most of my reservations about him had evaporated, and I thought him honest and direct, very unlike most of the double-talk artists in the agencies. We were both getting caught up in the morality of what we were doing as well, and reinforced each other about the importance of the case. If Rheingold had taken over the case, I probably would have quit—if not immediately, then somewhere along the way. And my instincts about him were perfectly right: later he quit PSC to become General Counsel for one of the biggest U.S. power companies.

Another problem, which developed because of the intra-PSC confusion, involved Matias. The power company lawyers were talking to him, trying to arrange the rules in a way that would best favor their position. Specifically they wanted me and Becker to testify first, so they could see what *we* were going to say before they filed their own testimony. Well, that burned me up. Already in the Rochester hearing there had been concealment and subterfuge by the companies—they had ignored Becker's letter, they had refused to release to Simpson certain secret studies done by the industry, and they had kept their own witness in the dark about damaging information—so why the hell should they get an advantage now? Such a deal would have relieved them of any burden to discuss evidence for health risks that they knew about; they would only have to react to what we said. Matias let it be known that he thought that was a fine idea. But we objected, strenuously, and eventually Matias changed his mind. Everybody would file testimony simultaneously.

Finally, Matias was badgering me, through Simpson, to renege on our initial request that hearing sessions that involved us should be held in Syracuse. Apparently he didn't want to have to travel away from Albany. The more I looked at Matias, the more petty he seemed. I really began to worry about testifying before him, and having him judge me.

*

As the hearing grew more complex, I felt myself being pushed farther and farther from the warm, comfortable, uncontroversial life I had led before Simpson came. I was married to a fine woman and had three young sons. I had a house in the country with a garden and five acres of land. I had a secure job in an exciting lab, and a boss who was going to be famous some day—and who treated me with generosity and confidence in my abilities as a scientist. It's true that I was still uncertain about my ultimate place in the scheme of things, about my overriding purpose in life—my Jesuit teachers at St. Joseph's had embedded in me a sense of ethical obligation that I still hadn't fully identified or understood—but otherwise things were going well. Basically I felt I had control of my life. But now, with the proliferating complexities of the hearing, I wasn't so sure anymore. I wanted to quit... to get off the spot...to go back to the way things were. I didn't want people angry at me, especially people who could do me harm. And even though Becker and I had obtained permission from the VA Director to testify, I was scared that the VA officials in Washington might decide that I was too controversial and maybe even fire me. At my lowest points I even worried that the Character Review Committee of the New York Bar would find me too controversial, and hence unfit to be licensed to practice law in New York.

Perhaps I lacked the courage to quit. I don't know. Maybe I was afraid it would look as if I was being pushed out, and what would everyone think? (By this time the press had begun to interview us.) But I really think it was something deeper that made me stay on. I viewed myself as a logical person, like Mr. Spock. Why should I quit? Nothing has happened, only vague fears and rumors. Besides, I'm right, I know I'm right. It simply wouldn't be logical to quit. So—I couldn't quit and be logical, and I couldn't stay and be content. In the end I stayed, and in the process of deciding that I learned quite a bit about myself.

*

Early in March 1975 Simpson called to tell us the names of the new experts who had been hired by the companies. One of the names on the list gave me a real start—Herman Schwan.

When I had been taking graduate courses in Philadelphia, before I came to Syracuse, I had audited one of Schwan's courses. He was so intimidating in the classroom that students rarely even asked him a question. Later, when I came to Syracuse for my Ph.D., Schwan's papers had often been required reading.

He was one of the most prominent biophysicists in America. Why, I wondered, was he testifying in a state powerline hearing? Surely he knew it would be time-consuming, politically controversial, a general pain in the neck? It seemed incongruous.

There was only one other name on the list that either of us recognized, and it was almost equally puzzling. It was a veterinarian named Solomon Michaelson, from the University of Rochester. Several years earlier Becker had hired Michaelson to inspect the VA's animal research facility, and the reason Becker remembered him was that someone high up in the VA had *told* Becker to hire *him*, rather than someone else; it seemed to him that Michaelson had some sort of clout. But Michaelson hadn't impressed Becker as anything more than a competent vet—why was *he* testifying in a powerline hearing on human health risks?

CHAPTER 4

The Biophysicist and the Veterinarian

Putting aside the question of whether it was wise, I was in a legal proceeding and I tried to think like a lawyer. The issue was the nature and extent of any health risk associated with exposure to the NIEMR of a powerline. How would the power-company experts testify? I looked for the answer in what they had said and done in the past.

Herman Schwan had come to America from Germany in 1947 under Project Paperclip, a controversial government program to import German scientists immediately following the end of World War II. (See Clarence Lasby's *Project Paperclip: German Scientists and the Cold War*, Atheneum, 1971.) Schwan worked for the U.S. Navy until 1950, when he became a professor at the University of Pennsylvania. He had done some kind of NIEMR research in Germany during the war, but I couldn't determine exactly what. After he came to the USA he began to publish papers which said that the laws of physics, properly interpreted, showed that the only effects of NIEMR on living things would be through heating or shock.

It seemed that Schwan's writing was intimately bound up with the federal government's concern, which surfaced in the 1950s, about military employees who were beginning to report various injuries from working around radar—eye injuries, temporary and permanent sterility, internal bleeding, and other problems. In response to these reports, an Air Force surgeon, Colonel George Knauf, was picked to determine how much NIEMR was safe for personnel. Knauf and Schwan began to work together. Although the USA was then a leader in NIEMR-producing technologies, its expertise was in engineering rather than the biological aspects of NIEMR. Of all the biological experts at that time, Schwan was the most important.

Schwan regarded the stories of nonthermal injuries—those which happened to people without any evidence of heat, without their having *felt* anything—as anecdotal and unreliable. Either these effects were secondary to actual heat, he reasoned, or they didn't exist. Thus, to Schwan, NIEMR was safe if it didn't cause heating. Well, then, what was the maximum level that would not cause thermal effects? Schwan's answer was simple. The body could handle a certain amount of heat, for example by sweating, but if the heat reached the point at which the body's defenses broke down, temperature would rise and injury would result. According to his calculations the safe level would be 10 milliwatts per square centimeter (a measure of power

14

intensity delivered to an area of surface). No heat under that level—and no effects without heat.

Schwan's numerical value apparently provided Knauf with just the handle he needed. The problem became a simple one of validating Schwan's thermal threshold. Even in the 1950s, there were some who felt that Schwan's position of heating-effects-only was not correct, but from the beginning their opposition never amounted to much, and it seemed that Knauf never seriously questioned Schwan's assertion that heating was the only possible effect.

In 1955 the Department of Defense provisorily accepted Schwan's safety level of 10 mW/cm2 and Knauf got the go-ahead to fund a series of animal experiments to verify Schwan's calculations. The studies were called the Tri-Service Program.

One of the participants in this program was Solomon Michaelson at the University of Rochester. Michaelson had become an expert in the use of the beagle dog as a test animal, and, in a revolting series of experiments, he literally cooked dogs alive with NIEMR at levels of 50 to 100 mW/cm2. He measured and recorded such things as the location of burns, the oozing of fluid from the brain and eyes, and body temperatures, which reached 106–108 F. In each case, Michaelson noted the exact cause of death which resulted when the dogs could no longer "handle" the heat. Later, in the 1960s, he did similar experiments on dogs with x rays, again at lethal levels.

Michaelson's work for the Tri-Service Program was basically confirmed by other investigators. Gross, acute effects had been observed at NIEMR levels only slightly above Schwan's safety level. From this finding it was concluded, with strange logic, that Schwan's level was safe. There was not one instance of a Tri-Service Program experiment conducted at intensities *below* Schwan's level.

Schwan was subsequently appointed chairman of a committee of the American National Standards Institute (ANSI), whose goal was to set a NIEMR level for industry. It came as no surprise when ANSI also accepted Schwan's position and 10 mW/cm2 became the "safe" level for such industries as radar and radio, as well as many that used NIEMR in the workplace to do a variety of jobs—sealing packages, treating seeds, operating equipment, providing security, and many other tasks.

Schwan published dozens of papers and gave hundreds of lectures over the next twenty years, all reflecting his view that heating was the only NIEMR effect. His rise was steady in the American scientific establishment, culminating in his election to the National Academy of Engineering.

It was hard to work through this complex history, but it seemed that there was general government and industry agreement that 10 mW/cm2 was safe—that is, that there were only thermal effects. And it also seemed that the scientific basis for this view rested heavily, maybe even exclusively, on the shoulders of Herman Schwan. Whenever anyone referred to the 10-mW/cm2 rule, Schwan was usually cited as the authority; and Schwan himself usually cited others only for support, not for authority.

What Schwan actually said in most of his papers was that there were no *known* biological effects of NIEMR below 10 mW/cm2. But in fact there were some such reports, particularly from the Soviet Union, whose existence Schwan never seemed to recognize. What Schwan seemed to convey was that there couldn't be any such effects. Certainly the common opinion was that 10 mW/cm2 was safe. But since almost no work had been done in the USA at or below that level, the only possible scientific basis for Schwan's view must have been a mathematical or theoretical analysis. Indeed, as I found, that was true. Schwan's view was based on calculations involving nonbiological models.

The NIEMR situation, and Schwan's role in it, was also disturbing for me in another way. In Becker's lab I had been doing experiments with bone, tendon, and other tissues, measuring their electrical and magnetic properties. There was very little government or industry interest in this work. I did an experiment, published a paper, and then did another experiment. The NIEMR exposure situation was just the opposite: I found twenty-two federal agencies that had some jurisdiction or interest in NIEMR bioeffects, and the number of companies—whole industries—that were interested was simply huge. What Schwan said or wrote was very much a matter of interest to them. He was a consultant to this company and had a grant from that agency—everything he did seemed to be related to companies and agencies, and to the military.

Despite his awesome reputation, Schwan did little actual experimental work himself, particularly after 1970. His work was mostly mathematical analysis, and the rehashing of his earlier theoretical calculations. I felt that Schwan's papers were not serious attempts to prove the validity of the 10-mW/cm2 rule, they were rationalizations of it.

All in all, Schwan's track record suggested that in the New York hearing he would do a calculation and conclude from it that the powerline would be safe. But calculations involving living organisms are a lot like statistics—with a little creativity one can make them say almost anything. Since Schwan wasn't doing any biological experiments, I wasn't too concerned about the substance of what he would say. Yet I was awfully wary about him personally and his connections—he looked like a big fly swatter, and I felt a little like the fly.

What about Michaelson? What was he going to say?

After his work for the Tri-Service Program, and beginning sometime around the late 1960s, Michaelson began to advocate strongly the position that, as long as NIEMR levels were below those he used on his dogs, they were completely safe. About 10 mW/cm2, he said, was a good figure. Above that the body couldn't handle the heat. Moreover, said Michaelson, the scattered reports by American, German, and Soviet scientists that NIEMR below the thermal threshold could cause effects were all incompetent and should be ignored. Michaelson was especially critical of the Soviet scientists who found effects. He accused them of hiding data, and intimated that they were fabricating results. The Soviets were excessively concerned with safety anyway, he said; if the USA adopted their safety levels—which were far more stringent than ours—the harm that would be done to industry and the military would outweigh any proposed public-health benefit.

At the same time he began to say such things publicly, Michaelson's career took off like a shot. He was appointed to committees of the National

16

Academy of Sciences, the World Health Organization, North Atlantic Treaty Organization, President's Office of Telecommunications Policy, Electric Power Research Institute, American National Standards Institute, Veterans Administration, National Institutes of Health, Walter Reed Army Institute of Research, and the Navy. In many instances, Michaelson appeared at hearings in tandem with Schwan. In more than fifty publications, he reiterated his position regarding NIEMR safety levels.

But in a Congressional hearing in 1973, Senator John Tunney of California brought Michaelson's attention to the fact that he had changed his position. Before 1967, Tunney pointed out, Michaelson had said that there were too many unknowns to try to establish a NIEMR safety level, that the 10-mW/cm2 standard had to be re-examined, and that Soviet reports showed there was a basic nonthermal effect. But testifying now, on behalf of the microwave oven industry, Michaelson was saying that 10 mW/cm2 was "exceedingly safe," that it was "inconceivable" that lower levels could be hazardous, that "no new data from the literature and no new valid arguments have been presented to change the situation from what it was last year, four years ago, or fifteen years ago."

"Have you changed your opinion?" asked Tunney. This was Michaelson's answer: he was now "affiliated with many organizations" and, as a result, "I feel more confident now than what I appeared to have been in 1967."

Michaelson had published some surprisingly nasty articles attacking the work of one particular American investigator, an ophthalmologist named Milton Zaret. These articles led me to Zaret, and to some interesting information. Zaret had worked on a research program called Project Pandora. This was a secret Defense Department research program set up in the mid-1960s to study the biological effects of NIEMR, and its apparent aim was to determine whether the Soviets were carrying out offensive "warfare" against the employees of the U.S. Embassy in Moscow by means of a NIEMR signal beamed into the building. Zaret had already found that nonthermal levels of NIEMR could cause cataracts, when he was asked in 1965 to become a medical investigator for Pandora. In the course of his research he duplicated some Czechoslovak experiments in which NIEMR bioeffects had been found, and performed other studies involving effects on the eyes. He wanted to continue his work, but in 1973 a committee headed by Michaelson told Zaret his work was worthless and recommended that his funding be terminated. After that Michaelson bird-dogged Zaret on the question of NIEMR-induced cataracts. Whenever Zaret reported his results, Michaelson—a veterinarian—would dispute them. In one instance Michaelson wrote directly to the editor of a journal that had accepted one of Zaret's papers, to try to have it withdrawn from publication. What made the whole story really incredible was that Zaret, at Michaelson's request, had operated on Michaelson's aged mother and had restored her sight, which had been lost because of cataracts in both eyes.

I had never seen anything like Michaelson and what he was doing. He was like a Manchurian Candidate aimed at the opposition to the 10-mW/cm2 rule. The thing that cinched it in my mind was his testimony for Rockland Utilities in New Jersey.

Rockland wanted to build a large antenna, and the local landowners were concerned about its safety. Michaelson testified on behalf of Rockland and said that no one would be exposed to levels above 10 mW/cm2, and therefore that the antenna was safe.

With Michaelson on the stand, a lawyer representing the citizens held up a book on NIEMR bioeffects written by Dr. Zinaida Gordon, a Soviet scientist. The book described experiments in which adverse effects had been found at nonthermal levels. "What do you make of this, Dr. Michaelson?" asked the lawyer. "Well," Michaelson said, "she's a woman."

When the lawyer recovered enough to ask Michaelson to explain what he meant by that, Michaelson began attacking Dr. Gordon's research as faulty. One of the biggest flaws, he said, was that she had used metal cages for the animals, when she should have used plastic.

Question: What is polystyrene?
Michaelson: That's a plastic.
Question: (quoted from Dr. Gordon's book) "... the cages were made of polystyrene...." Now, Doctor, do you still state that this woman used steel cages?
Michaelson: Yes....
Question: In other words, even though she told everybody she uses polystyrene, she uses metal cages, is that right?
Michaelson: Yes.
Judge: Is it possible that she had the wrong picture in the book?
Michaelson: No....

There was no doubt in my mind what Michaelson would say in the hearing: he would attack anyone—read Becker and me—who reported bioeffects due to NIEMR.

The power companies also had other witnesses who like Michaelson were from the University of Rochester. None seemed to have any relationship to the NIEMR issue, so there was no possible way I could anticipate what they would say. The one exception was a botanist named Morton Miller. Miller was a Sanguine investigator who had studied the effect of Sanguine NIEMR on bean roots. I presumed that he would testify that the powerline NIEMR would not hurt the grass and trees. Since the hearing was about possible health risks to people, I didn't think that was very important, so I never gave Miller much thought.

As a lawyer, I didn't think the power companies had a significant affirmative case. All they could do would be to present calculations, call people names, and perhaps show that the grass and trees would be okay. Now the law says that the power companies have to show that the line will be safe before they get a license—that is, they have the burden of proof on the issue. If they can't sustain that burden—and I didn't see how they could—then they don't get the license. I talked to Simpson about this point from time to time in 1975, and he told me that Matias didn't agree. Matias thought that since Becker and I had raised the issue, the burden of proof was on us. Well, I thought that maybe he was just jerking us around with comments like that because we wouldn't agree to have the hearing in Albany. But it turned out that Matias was more serious than I thought.

CHAPTER 5

Further In

Throughout 1975 my day-to-day goal was evidence—I wanted to prove beyond question that I was right. But I also wanted to see where it would all lead. I talked to a lot of different people, and it was hard to predict who would be friendly and forthcoming, and who would be indifferent, or hostile. Often, it seemed, people had special interests to protect—a contract, a grant, a company, a government agency. Promising leads would dead-end because someone would fail to return a phone call or refuse to send a copy of some report. During this time, I lost the last lingering fear that I was wrong.

Nineteen seventy-five was a pivotal year for me in another respect. Until then I had worked closely with Becker, but thereafter we began to separate. More and more Becker began to cover for me: he dealt with various bureaucratic levels within the VA to maintain the laboratory environment that permitted me to function. Frequently he needled me that I had all the fun and he had all the paperwork. It was true, but he allowed it to be so. I always thought that, on balance, he thought what I was doing was worth his grief.

Normally, a lab director would use someone such as me to help carry out focused experiments to try and prove his pet theories. Indeed that's what I had done for Becker in the past. Such activities lead to papers, which lead to grants, which provide money to continue the cycle; but when a director uses his resources or something else besides papers, it tends to jeopardize his future because it slows the growth of his scientific stature. Nevertheless, Becker gave me my head. He is a highly ethical person and has a great compassion for people—ordinary people, not authority figures, whom he seems generally to dislike (most lawyers, for example). He is also the most stubborn man I have ever known—in sixteen years I don't think I ever saw him change his mind. The way I see it now is that even though not all the specific facts were in hand—"every jot and tittle," as he would say—Becker believed that there was a health risk from the powerline NIEMR, and that the constant exposure of the public to it was wrong. In response he donated me, for whatever good I might do, and lived with the diminution of his assets and the consequences that befell him because of my activities.

Becker did not take part in the day-to-day developments. The effort required many calls, visits, letters, and library searches, and he had neither the time nor the inclination to engage in such activities. He was the philosopher and generalist, and the task at hand—the obtaining of evidence—called for a technician.

19

One trail of evidence I pursued involved the power companies themselves and their actions regarding NIEMR.

Sometime in the mid-1960s the industry had apparently obtained copies of Soviet studies on the biological effects of powerline NIEMR. Among them were reports describing various illnesses in power-company workers that arose from prolonged exposure. Without openly acknowledging its motivation, the American Electric Power Company (AEP), one of the largest power companies in the USA, commissioned a two-pronged study by scientists at Johns Hopkins University, the results of which were released in 1967. In a survey of eleven linemen, the investigators found them to be generally healthy, except that two men had reduced sperm counts. A second study found that exposed mice were not harmed by NIEMR, but that their offspring—which were not themselves exposed—were stunted. I learned from one of the Hopkins investigators that the team had requested funds for further studies and had been turned down. During the next several years, the industry emphasized that the linemen were healthy and argued that therefore the general public, which would be exposed to much lower levels, was in no danger.

There was no further significant activity until August 1972, when the situation flared up again. At an international conference on high-voltage powerlines in Paris (CIGRE, from the French name), Soviet engineers announced for the first time directly to the West that they had performed studies of the effects of powerline NIEMR on workers and found that they needed protective clothing. Without it, they said, "Young men complained of reduced sexual potency," and there was an adverse elect on "the central nervous system [and the] heart and blood-vessel system." The longer a worker was exposed, they said, the worse the symptoms.

The Soviet investigators went on to describe certain nationwide rules governing permissible exposure durations. After they had finished, the chairman of the meeting asked what observations had been made in other countries. Howard Barnes, an engineer for AEP who had worked on the Hopkins studies, registered strenuous disagreement with the Soviets, and said there had been no complaints of NIEMR-caused health effects in America.

Following CIGRE, Barnes wrote the Soviet colleagues for copies of the scientific studies they had used in coming to their conclusions, and over the next few years there was further correspondence. The Soviets supplied Russian-language copies of the reports that had led them to adopt NIEMR work rules, and at the same time asked whether the USA had similar exposure standards. Barnes replied that there were rules governing the height of powerlines above the ground for "the safety of firemen fighting building fires." But that was *not* what the Soviets were asking for, and they wrote back: "Are there any rules for exposure to [NIEMR], and if not, are they being developed?" The answer to both questions was no.

The industry released translations of the Soviet reports. They showed that the Soviets had studied several hundred linemen, not only eleven as the Americans had done. Also, the Americans had performed only standard physical examinations; the Soviets had done psychological and neurological testing. It was these tests that revealed the harm done by NIEMR. The released translations were preceded by an introduction, written by Barnes, which emphasized that the Soviet results were at variance with the 1967 studies done by AEP. The reader was presented with the notion of a conflict between rock-solid American science and the results of Communists.

The introduction of the booklet on the Soviet studies invoked a particular industry argument for powerline safety that I was increasingly encountering. I called it the Operating Experience Argument: There were 500,000 miles of high-voltage lines in the USA, and there wasn't a single report, not one confirmed case, of anyone being killed or made ill by the NIEMR of such lines, so they *must* be safe. The premise was true but the conclusion was false. There had been *no studies* of the side effects of living or working near such lines, so it was hardly surprising that there were no proven cases of actual harm.

*

I was admitted to the New York Bar in mid-1975, an event that caused me to reflect about the issues being raised as the hearing approached. Lawyers are cautioned against imputing sinister motives to their adversaries because the aim of the legal process is to purify and refine the issues and frame them for decision. Because it is adversarial in nature, the legal process inherently involves a polarization of views—the creation of a black-and-white world where the subtle grays of reality frequently don't show up. I knew and accepted this situation up to a point, yet it was becoming harder and harder for me to ignore motives and ethical considerations in the NIEMR issue. After I came across Cyril Comar and his friends it became impossible.

One day I got a phone call from an engineer who was working on a powerline NIEMR study at Johns Hopkins Medical School. He said that he had read Becker's and my 1974 testimony, and he wanted us to know that he and others were doing related research. In one study, they found that NIEMR killed cells exposed in little plastic dishes in less than one week. The investigators, headed by Dr. Donald Gann, then began a second study involving the effects of five-hour exposures of dogs to NIEMR. The results were startling: despite the brief exposure period, the dogs exhibited changes in blood pressure and heart rate. In his June 1975 monthly report to Cyril Comar at the Electric Power Research Institute (EPRI), which supported the study, Gann said:

> The unexpected findings of these changes suggest strongly that dynamic effects resulting from exposure to [NIEMR] may not be particularly subtle at all, but may be quite easy to detect. In addition to the findings with respect to magnitudes of change, the variability in the heart rate of exposed subjects was also significantly greater than that in unexposed subjects, suggesting that the observations made by Soviet workers on conscious human beings exposed to high-voltage [NIEMR] may be present in anesthetized dogs. These results are clearly preliminary but also clearly demand a further exploration.

At the time, this was the only work in the USA on powerline NIEMR safety. It was being done at a prestigious institution by investigators chosen

21

by EPRI, and I thought that EPRI had a clear and obvious duty to divulge the results, and to continue the study. It was the first indication of possible harm that had emerged from within the industry itself and it therefore could not be honestly dismissed as the work of Communists or incompetent scientists.

Instead, Gann's work, which was a vulnerable seed of information that needed to be nourished, fell on utterly barren ground. In June Comar wrote to Gann that a review committee—which included Michaelson—had concluded that his data gave "little basis for follow-up study," and that further funding by EPRI would be "completely unwarranted." Comar said that EPRI believed the work "lacked any appearance of accurate scientific reporting," was "substandard," and was characterized by a "lack of professionalism." Despite this attack on Gann's work, Comar said "This decision is in no way a reflection on your professional competence." The study was cancelled, the research team scattered, and the study animals were destroyed. Even though the effort had cost over $500,000, neither its existence nor the tentative results were acknowledged by the power industry.

EPRI'S main business was developing new technology. In 1975, at a test facility in Pittsfield, Massachusetts, it was already testing equipment destined for the next generation of high-voltage lines, which would operate at 1.1 million volts. I visited the Pittsfield facility in June 1975 to learn more about these lines, and the experience accentuated my developing perception of power-company officialdom as insensitive to the biological consequences of what they were building.

Pittsfield was, first and most obviously, a place for engineers. There were erector-set towers of breathtaking size, cables as thick as one's wrists, insulators so big that it took three men to lift them. Standing under the test line I felt strong emotions: awe of the technological feats man is capable of, the sense of man's apparent destiny to build such complex things, the sense that somehow it had to be the right thing to do. But every place I looked I saw something to do with engineering—measurements of how strong this was, or how much of that was being emitted. It all seemed trivial compared to the questions of what would happen to people and the environment once the thing was built and they were exposed to the things that would be emitted. Perhaps these questions were so difficult to answer that a decision had been made not even to try.

With the exception of a pathetic patch of corn someone had planted under the line, nothing was happening at Pittsfield that had anything to do with effects on living systems. Yet every instinct said there had to be effects. The very act of walking under the line was painful because of the little shocks that occurred when blades of grass—constantly charged by the line's NIEMR—brushed against one's ankles. To force people to live in a similar environment for their whole life was not just wrong, it was heinous. It burdened me that science had made it possible.

In talking to the engineers there, I learned that some of them had developed problems with their joints about the same time that the line had been energized. I told them that the orthopedic surgeons in the research group at the VA would be glad to examine them, and to compare their pre- and post-exposure x-rays. I was later told that EPRI considered such assistance "inadvisable."

In 1975 the U.S. Government made an effort to bridge the gap between

U.S. and Soviet views on the powerline NIEMR issue. Engineers from both sides met in Washington under the auspices of the Department of the Interior. It was plain from the beginning that things were no different from the way they had been at the CIGRE conference three years before. While the Soviets were instituting a major research effort and developing more stringent safety levels, the Americans stoutly maintained that there was no need for any safety level because the NIEMR was completely harmless. Privately, some of the U.S. engineers felt that the Soviets had trumped up the whole thing just to irritate them and cause trouble.

I couldn't help concluding that the U.S. power industry had at best a foot-dragging hear-no-evil, see-no-evil approach to a problem that was going to be very expensive to deal with. At worst, I thought, the industry just didn't give a damn about the human beings who were being exposed because, whatever the NIEMR might be doing, it wasn't obvious, so the companies couldn't be sued.

The power industry position was, so far as I could see, basically rooted in financial considerations. If they could have built their lines as cheaply without exposing people to NIEMR, I think they probably would have done so. But our testimony in New York was a direct challenge to the thermal standard for *all* NIEMR, and it was the military whose interests would be most crucially at stake if such a challenge succeeded. What bothered me most was the possibility that Schwan and Michaelson, who had been supported for many years with DoD funding, had joined the hearing at the behest of someone in the military. I had no evidence for this connection, yet there were developments during 1975 which further suggested deep concern, especially on the part of the Navy, about the developing evidence for a health risk from NIEMR.

*

By mid-1975 almost all the information to which Becker had been made privy at the 1973 Sanguine meeting—except the committee report itself— had been released. There were two projects that particularly interested me. The first was done by Dietrich Beischer, who was director of the Naval Aerospace Medical Research Laboratory in Pensacola, Florida. Beischer found that human volunteers exposed to Sanguine NIEMR developed elevated levels of blood triglycerides, seemingly confirming an earlier observation made at a test facility in Wisconsin. The other project, which was carried out at the Naval Air Development Center, Johnsville, Pennsylvania, had involved exposure of rats to weak NIEMR, and had apparently found stunted growth.

Becker had known Dietrich Beischer since 1962, when Beischer had invited him to his lab in Maryland to observe the set-up for some experiments on magnetic fields and human biorhythms. Becker had been impressed with Beischer's care as an investigator, and over the years they had corresponded and had seen each other at conferences. Beischer's importance as a researcher had risen steadily, and in 1975 he was one of the world's foremost authorities on magnetic-field bioeffects.

Beischer, a contemporary of Schwan's, had also come to America from Germany via Project Paperclip after the war, but he was directly opposite to Schwan in demeanor. I wrote him and spoke with him on the phone, and found him modest, courteous, and helpful. In the early 1970s, as part of an effort to secure funds for Sanguine experiments at his lab in Pensacola, Beischer had written:

Besides serving the needs of an environmental statement by the Navy, the study may shed light on previously unrecognized aspects of utility power. Human partici-pants will be exposed exclusively to conditions under which millions of people live all over the world. However, the laboratory environment allows [one] to control the environmental conditions and to make tests which are expected to reveal subtle changes in the clinical-psychological-physiological makeup of exposed persons. Thus, the services of the few may benefit the public in general and the specific pur-poses of the Navy.

In his study of the Sanguine NIEMR, Beischer found that an exposure of 24 hours was correlated with elevated levels of triglycerides in 90% of the subjects. Beischer also measured other parameters but found they weren't affected. In a carefully worded conclusion, he said: "The present series of experiments should be considered as the beginning of an extensive research effort to find and characterize possible physiological and psychological ef-fects of [NIEMR] on man.... Barring the oversight of a crucial factor, the results of the present study strongly indicate that certain mechanisms of lipid management in the human body are influenced by an external, com-paratively weak [NIEMR]."

Beischer's report was a joyful thing to Becker and me because it was good evidence, and because it showed Beischer to be an intellectual ally.

Then something odd happened.

Becker had set up a meeting in Syracuse for scientific experts in mag-netic fields, and had invited Beischer to attend. Beischer had accepted. But a few days before the meeting there was a phone call. Becker listened as a voice at the other end said, "Hello, Dr. Becker, this is Dietrich Beischer. I cannot come to the meeting. They will not let me. I have to follow orders, I'm sure you understand." Then Beischer hung up. Becker sat there with a dead phone in his hand, stunned.

In March 1975 Beischer's name appeared as first author on Navy report NAMRL 1197, which described an experiment actually performed by oth-ers. It failed to find any triglyceride effects in exposed mice, and was writ-ten in such a way as to cast doubt on Beischer's work with humans.

Soon after the phone conversation, we learned that Beischer had retired and was living in seclusion in North Carolina. To my knowledge, he has neither written nor spoken about NIEMR bioeffects since that time.

*

In my exploration of the other study, the Johnsville investigation that had apparently turned up stunting effects in rats, I found a brief description under the name of "Kendrick" as principal investigator. I called Johnsville, and spoke to Personnel and the base librarian, but was told there was no record of any report written by anyone named Kendrick, nor any evidence Kendrick had ever been there. I had been doing rat studies and getting ef-fects, and I wanted to know the details of what Kendrick had found. But it was as though the study had disappeared.

I did find one report which referred to rat studies at Johnsville. That was the first time I came across the name of Philip Handler, then president of the National Academy of Sciences (NAS).

NAS has some of the most famous scientists in the USA as members. Election to NAS is considered very prestigious, yet such an honor is essentially ceremonial because NAS itself has few important functions. The National Research Council (NRC) is the operating arm of NAS and actually carries out its scientific and political functions. The NAS president is the head of NRC; and it is NRC, with its large staff and its facilities, which is the significant source of the president's clout. When NAS gives a scientific opinion about this or that, the job has actually been carried out by NRC, or more precisely by a committee appointed by the NRC president. Members of such ad hoc committees need not be, and usually are not, members of NAS. The theory is that the best people are supposed to be chosen for the particular task, regardless of whether they happen to be NAS members.

The NRC report I obtained which referred to Johnsville was in form a review of some Navy NIEMR projects. By and large it was superficial and nonspecific and did not evince any strongly held views. But when it referred to the rat studies at Johnsville, its tone was very different. The committee was "shocked" to see the poor quality of the work that had been done there, and it felt that the study couldn't possibly have any value. The NRC report was dated 1974 and referred to the Johnsville studies in the past tense, so I inferred that they had already ended. Otherwise the report contained no hard facts about the Johnsville project beyond the little I already knew via Becker's 1973 meeting. But whatever the results, they must have been simply amazing, because the report's authors—Michaelson and Schwan—registered their highest indignation level and strongest condemnation of any NIEMR study up to that time.

I wanted the facts about the study because it seemed that they would be good evidence for my argument. But it was after all only one study, and since I had others I didn't regard the gap created by my inability to obtain it as crucial. What really fascinated me was who at NRC had appointed Schwan and Michaelson to review the Navy's NIEMR work? The answer was that the responsibility for picking each member of each NRC committee ultimately rested with the NAS president.

I had never heard of Handler, but Becker had, and he disliked him. He viewed Handler—a biochemist—as a strong proponent of orthodox, not to say safe, science. To Becker, who was nothing if not innovative, it was unfortunate that Handler occupied the most prestigious public leadership position in the U.S. science establishment, for it meant among other things that he had the power to obstruct any developments in science if he so chose.

I approached the question of why Handler picked Michaelson and Schwan for the Johnsville review committee through the conflict-of-interest rule, which I thought should have alerted Handler to their potential bias. But the NRC staff was very uncooperative, and I learned little, except that "bias statements" are not required from individual committee members until after they have been chosen. I also found that NRC never releases bias statements to the public, is in fact not required to do so since it isn't covered by the Freedom of Information laws, even though NAS is a corporation chartered by Congress. So my inquiry into the facts of the Johnsville studies met a stone wall.

*

Initially, it seemed to me, Becker hadn't been especially bothered by what my search into the NIEMR issue was uncovering. He was intimate with scientific conflict; he often said that new ideas in science usually succeed less because of their merit than because those who oppose them eventually die. But now he could see it was less a bona fide scientific dispute than political intrigue in which we were becoming enmeshed. The phone call from Beischer and its aftermath really bothered him, and he began to be concerned about the risks we ran in testifying. He understood better than I how anonymous pressure could be brought to bear to make life very difficult for us.

But aside from the personal worry, there was the risk to his research. As long as those in control of the NIEMR health-risk issue denied that nonthermal effects of NIEMR on human beings existed, it seemed likely that advances in the therapeutic use of NIEMR would be retarded. The century's explosive development of NIEMR technologies had occurred in a vacuum of concern over any medical implications. It seemed that if government and industry ever admitted that NIEMR had potentially *beneficial* nonthermal medical applications, they would *also* have to accept the possibility of human health hazards. What really angered Becker, then, was the possible fallout on important medical research from what people like Michaelson might say in a powerline hearing.

And there was another concern as well. It seemed likely to Becker that both the Soviet and the U.S. military establishments were involved in weapons applications of NIEMR. One day in the summer of 1975 he told me about a visit he'd had from the Central Intelligence Agency (CIA).

A CIA agent came to Becker's lab in 1970 to tell him about the Moscow signal and to ask his advice about its significance. Low levels of NIEMR were being beamed against the U.S. Embassy in Moscow from transmitters in and on the roofs of adjacent buildings, and the government was concerned that it might be harmful to Embassy employees. The agent wanted to know whether Becker thought it might be a health hazard. Becker said yes, he thought that was possible. The agent went on to tell Becker about the government's suspicion that the Soviets could alter personality without leaving a trace of evidence of how it was done. He mentioned American flight crews which had been captured by the Russians near the Turkish border, and said when the men returned to the U.S. they seemed very different though they said nothing in debriefing sessions about any sort of intimidation, torture, or harassment. Something, the government reasoned, had been done to them without their knowledge, and this something intensified suspicions that the Moscow signal was an offensive act.

The agent asked if Becker would pass along scientific literature that might be useful, and Becker said he would. After that meeting he had sent out occasional copies of papers dealing with NIEMR bioeffects. He didn't like the business much, and he kept quiet about it. His decision to tell me about it was a matter of self-defense; the more I knew as I explored the issue, he figured, the better off we both would be.

PART II

CHAPTER 6

Preparations for Battle

The scientific issues had never been examined in a court of law before, so as far as I was concerned everything came down to the hearing in New York. As complex and far reaching as the NIEMR issue seemed to be, there was only one way Becker and I could have an effect, and that was to beat the power companies soundly in New York. A win in that forum, where the assumptions, evidence, and the scientists themselves would be subjected to the most extensive examinations, might open the way to a serious national re-evaluation of the whole NIEMR problem.

The contest was shaping up like this: Schwan would probably perform his calculations to show the powerline was safe; Michaelson would no doubt attack the researchers who had found effects; Miller would say the line couldn't harm plant life. Perhaps one of them would drag in the "operating experience" argument as well—probably Michaelson. Becker and I would present the results of our own experiments and those of others, and conclude that the sheer number of reports of diverse NIEMR bioeffects meant that the line would be an unacceptable medical risk to humans. I thought about revealing what I knew of Michaelson's past, but decided that was best left to Simpson during cross-examination; otherwise it would have diverted attention from the issues in my own testimony.

I had also thought at one point about countering Schwan's calculations with my own, but I decided against it. To do that would turn the hearing record into a physics textbook, which would be incomprehensible to the commissioners. Moreover, a battle of calculations would likely favor Schwan and the power companies because, in the resulting confusion, those in favor of the status quo would have the advantage. The burden was squarely on us to prove the need for change and for new regulations, and the only way that could be done, I thought, would be to engulf Schwan's theories in a torrent of experimental facts. In regard to Michaelson, too, the more studies cited the better, because it would put him in the position of having to attack them all, and at some point PSC would realize he was not a credible witness.

The facts that I had discovered in 1975 from scouring the literature were overwhelming. Aside from our own studies, I cited thirty others that described the impact of powerline-type NIEMR on a wide variety of life pro-

27

cesses. In humans, it appeared to alter reaction time, triglyceride levels, psychological performance, and biorhythms; in rats and mice it affected electroencephalograms, blood cells, growth rate, and enzyme levels; it produced a diverse range of alterations in chickens, brain cells, amoebae, birds, worms, slime mold, bees, dogs, and monkeys. In some of the studies the strength of the NIEMR corresponded to that which would occur several thousand feet from the powerline.

The first of our own experiments had involved preliminary measurements on rats exposed for thirty days to a NIEMR level almost identical to that under the wires of the 765,000-volt line. I had done the experiment three times in 1974, and repeated it seven more times in 1975, each time with some improvement in the procedure. Both the body weights and the levels of various blood chemicals were affected, and at the end it was clear that the NIEMR had produced biological stress in the animals.

Our second experiment was an attempt to see what the NIEMR would do to a biological system over several generations. As far as I knew, such a study had never been done before. But if people were going to be continuously exposed to the powerline, and then raise their children in the same environment, it made sense to try to determine what the effects over several generations would be like in a lab animal.

I designed and built an apparatus that would expose mice to NIEMR from the moment of birth through maturity and mating. The offspring of the first generation would experience the same thing, and then their offspring as well. Three generations would be exposed. The experiment took six months to complete. One day, toward the end, Becker went to the animal care facility to see the mice. The results were truly startling. One set of cages contained adult mice of normal size; in the other set were adult mice about half that size. The former were unexposed controls; the latter, the third generation of exposed mice.

Becker told me to take pictures. A photo, he thought, would hit PSC much harder than a string of numbers. How could they ignore the contradiction between Schwan's calculations and photos that *showed* those calculations were wrong?

In Schwan's thermal-effects-only view, such effects were literally impossible. Indeed the results of all thirty-two experiments were impossible. So the issue, as I saw it, was simple and straightforward: either Schwan was wrong, or thirty-one research groups were wrong.

Although the evidence was clear, the problem in a legal forum was to present it in such a way that the power companies' lawyers could not effectively challenge it. Part of the difficulty arose from having to extrapolate from effects seen in animals, under a variety of controlled lab conditions, to the certainty of effects in humans living near the lines. I had to anticipate that the company lawyers would focus on the uncertainty of effects in humans. How do you *know* there will be definite effects in humans? they would ask.

For me to say only "there will be a risk" would seem weak and subjective. To say "there will be biological effects" would provoke the lawyers to ask "What effects, specifically?" and "Who will suffer them?" To specify diseases would be impossible, a mere guess. To say I didn't know who would suffer ill effects, or what they would be, would simply encourage the lawyers to argue that this was the same thing as saying I couldn't be sure there would be effects at all. It was very slippery ground. To a lay person looking at the evidence, the necessary conclusions might seem inescapable. But the hearing was going to be no parlor discussion. The power companies had big interests to protect, and PSC would not lightly recommend new regulations on the basis of amorphous conclusions. The logic of the argument had to be airtight.

Seeking the words that couldn't be electively challenged, I realized in the end that I could say no more than "probably." There *were* effects in animals, and many at NIEMR strengths well below what would occur near the powerlines, but they were *only* laboratory effects. Not only were humans a different species, but the conditions of their exposure would likely be quite different from any of the controlled conditions in the lab studies. People might leave the area for a while, or wear protective clothing, or vary their diets, or live in houses that were shielded in some way. One simply couldn't predict actual exposure conditions at all. The other factor that complicated my task was that the mechanisms which created the effects were not known. There was no universally accepted principle or theory that could be used to *predict* effects, and all I could legitimately say was that there were such mechanisms, probably more than one.

So my conclusion was this: "Exposure to the [NIEMR] of the proposed line will probably cause biological effects in some individuals." 'Probably' meant 'more likely than not'; it was the best I could do and not expose myself to a trap during cross-examination.

Becker's attitude was different. He didn't like lawyers and it was very uncomfortable for him to think that they would play a large role in a major scientific debate. Simpson had once told him that a NIEMR researcher in California, W. Ross Adey, had refused PSC's invitation to testify with the heated assertion that the issue was far too complicated for lawyers to handle, and that *he* would have no part in it. Knowing the man, Becker suspected that this response was partly an excuse not to expose his views in public; even so, he had some sympathy for that view.

It wasn't so much that Becker scorned the need to make his testimony legally unassailable as it was that he believed the issues were absolutely clear, and that our evidence was undeniable by reasonable men and women. And as far as actual regulatory action was concerned, there were precedents aplenty for government decisions to regulate pollutants solely on the basis of laboratory evidence. It seemed highly improbable to him that PSC could fail to order new regulations, regardless of the legal technicalities which the power-company lawyers might use to obstruct the truth.

Becker's principal role in the hearing was that of a medical doctor, and since the companies were not sponsoring a physician on their side, he felt his position was doubly strong. He would medically interpret all the reports that I would enter into evidence and elucidate the medical meaning of our own experiments on the rats and mice.

In his testimony Becker wrote that the biological effects of NIEMR demonstrated by the thirty-two reports were of two kinds: growth and functional. Effects on growth depended on several factors including the state of health of the subject and the conditions of exposure. Functional effects included changes in the way the organism worked. They could be strictly behavioral, such as alterations in response time; they could be changes measured by the abnormal production of biochemical substances; or they could be changes in biorhythms. Furthermore, Becker wrote, it was entirely possible that in studies that had reported only functional changes, there might have been pathological changes as well—unobserved because the studies weren't designed to observe them.

As to our rat experiments, Becker concluded that the animals had suffered stress. Chronic stress had long been linked to whole complexes of diseases—cardiac, renal, gastrointestinal, nervous—and furthermore, he wrote, "chronic stress results in exacerbation of any pre-existing pathological processes." As far as Becker knew, there was no essential difference between the way rats and humans responded to stress.

Becker reiterated his argument against experimentation without informed consent, and he closed with a discussion of recent scientific work on the relation between biological functioning and naturally present NIEMR. He described his own work on the body's electrical growth-control system, and discussed the possibility that the course of human evolution was intimately linked to natural NIEMR. The production of man-made NIEMR, he insisted, would be likely not only to produce stress, it might also interfere with such crucial human functions as memory, the perception of pain, and self-healing. All this material was necessary, he felt, to indicate the potential scope of this problem.

We considered these new documents far the ones we had prepared back in 1974, and in December 1975 PSC sent them to the power companies. Shortly afterwards, we received the testimony of Michaelson, Schwan, and Miller; it was as extreme and strident—and scientifically weak—as I had guessed.

Schwan presented the same calculations he had published twenty years earlier. He represented human beings as dielectric balls, and said the amount of powerline NIEMR penetrating the "human being" was so small as to be perfectly safe. The only thing it could do was cause heating or shock, and since his calculations showed that the amount of NIEMR created by the line was too weak to do either, exposure—regardless how close the "human being" was to the wires—couldn't possibly cause any effects. He said that if he were wrong, then "life, as we know it, would probably never have developed," and that there was little more to be learned about biological effects of NIEMR. He dismissed out of hand the experimental reports of effects. For all of this material, there was not one single reference to anyone else's work...it was as though Schwan had figured it all out by himself from first principles.

Michaelson's testimony was largely devoted to attacking the competence of Dietrich Beischer and the Soviets. He said Beischer's experimental design made it impossible to show a cause-and-effect relationship between NIEMR and elevated triglyceride levels in humans. Playing up the differences in Soviet and U.S. scientific practice, Michaelson said the Russian work had to be "viewed with caution," and that an "inordinate" significance had been attached to it. He accused the Soviets of hiding from the Americans, and suggested that the results of their studies might simply mean that

30

the workers were dissatisfied with their jobs. There was no reason to think the line would not be perfectly safe.

The only surprise was the testimony of Morton Miller. Emphasizing his academic title of "radiation biologist," the botanist Miller said that his role was to present analysis of the Sanguine experiments—both plant and animal. But he described only a few which, like his own, had not found effects. Those which *had* found effects, he simply listed, as though he expected that somehow they wouldn't become an issue in the hearing. He said he had personally visited a 765,000-volt line and found that the plants nearby were normal. Miller's testimony was the kind that makes cross-examiners pant with anticipation.

None of these documents contained a single reference to my experiments on rats and mice, and none of them paid more than cursory attention to the majority of the other thirty studies I had found in the literature. Most of those studies, in fact, were ignored altogether. The new power-company testimony was hardly more than a dressed-up version of the original testimony Simpson had shown us when he had come to the lab that day in 1974—the same conclusions, delivered by men with more impressive résumés.

Other witnesses also prefiled testimony, some for and some against the power companies. The most notable was that of Allan Frey, who was one of the first American scientists to point to possible health risks from NIEMR. His stand had from time to time brought him into sharp conflict with Schwan and (particularly) Michaelson. In his testimony, Frey cautioned against increasing powerline voltages without first studying the possible health risks. (For a discussion of Frey's earlier role see Paul Brodeur, *The Zapping of America: Microwaves, Their Deadly Risk, and the Cover-up*, New York: W. W. Norton, 1977.)

Surveying the respective cases, I felt highly confident. The only uncertainty now was the process itself. I remembered a quotation from Francis Bacon that seemed to warn against inordinate hope: "What is Truth? said jesting Pilate; and would not stay for an answer."

CHAPTER 7

Flak

In December 1975, just as the testimony was filed, the Sanguine issue unexpectedly flared up when Senator Gaylord Nelson of Wisconsin acquired a copy of the Proceedings of the December 1973 Sanguine meeting. Nelson, who had been responsible for obtaining the funding for the Navy's studies that had led to the report, knew nothing about the report itself, and he was furious to realize now that the Navy had withheld it. For two years, he said, Navy admirals had been coming by his office to tell him that there was no problem.

After reading the report, Nelson phoned several of the Committee members, including Becker, to find out whether the recommended follow-up studies had been done, and then he held a press conference and made the report public. The Navy, he said, had deliberately suppressed evidence of a potential health risk to the citizens of his state, and none of the Committee members he talked to knew about any follow-up experiments.

Within a few weeks of this development came a major announcement: NAS had appointed a committee of experts to conduct a full-scale inquiry into the entire scientific literature on Sanguine-type NIEMR. To my surprise and disgust, the list of members was headed by Schwan, Michaelson, and Miller. Furthermore, everybody else on the Committee except W. Ross Adey was from outside the field of NIEMR bioeffects, and some of them worked for NIEMR-producing industries. If ever there was a rigged committee this was it—I could have written its final conclusions five minutes after I learned of its composition. Philip Handler had, in effect, reappointed his 1974 Johnsville committee with a few additional people to make it look legitimate. But this time I felt he *knew* he wouldn't get an impartial decision because, less than two months before he appointed them, Schwan, Michaelson, and Miller had given a ringing endorsement of the safety of powerline NIEMR in their PSC testimony. Since Sanguine NIEMR was perhaps a million times weaker than that from powerlines, what could they possibly say about Sanguine NIEMR except that it would be completely safe?

This was a very bad time for me. I guess I had a kind of faith, back then, that government decisions involving science and the public interest were usually on the up-and-up, even if they weren't always correct. I figured that when wrong decisions were made, someone would come along and publish new information to correct the error, and the decision would be changed;

32

but I also figured that rarely happened because the blue-ribbon panel of experts that made the decision in the first place was unlikely to err. Well, now I know that was pure baloney. Handler's committee was going to rubber-stamp Sanguine, and in the process make me and Becker look like asses, and all we could do was send a letter and bitch about its composition.

I did try one thing. I called the man who had been made chairman of the committee, J. Woodland Hastings, chairman of the Biology Department at Harvard. He sounded like a nice fellow, but I could tell he didn't know a thing about NIEMR. And he said he just assumed that everybody on the Committee was an unbiased expert because "that's the way the NAS works." When I told him about Schwan, Michaelson, and Miller, he said he was "shocked." He said that he would call the NRC staff officer assigned to the Committee and try to confirm what I had told him. "If it's true," he said, "then either Schwan, Michaelson, and Miller are off the Committee and you and Becker are on it, or I'll resign." I think he did make an honest effort, at least at first. A few weeks after I spoke to him he wrote to me: "I am desperately trying to get your appointment." But I think Handler had other ideas, and Hastings did not succeed. By the end of March, Hastings had ended all contact with me—he refused to answer my letters or calls—and he remained as the committee chairman.

*

Meanwhile, in New York, other pressures were mounting. PASNY, in particular, was leaning heavily on PSC for a favorable decision. The private power companies such as RG&E and NiMo were fully subject to PSC's rules and regulations, but PASNY was different. It was a state agency, and it had great freedom of action including a blanket exemption from almost all the rules and regulations that applied to private power companies, *including* the requirement to demonstrate public need. Practically the only exception was that PASNY had to prove its projects wouldn't harm the environment or injure people.

PASNY was a powerful force in New York, generating over 15% of all the power and providing cheap electricity to industry. It issued tax-free bonds to finance its projects, and its property was exempt from taxes. PASNY was at the beck and call of Governor Hugh Carey, who routinely placed his close friends and confidants in positions of authority there. That gave Carey tight control over enormous quantities of cheap electrical power, which he used as an inducement to entice industry back into the northeast.

PASNY was onto a veritable goldmine of electricity in Quebec, where a provincial power company, Quebec Hydro, was building one of the largest hydroelectric stations in the world. The James Bay Project, involving the flooding of 60,000 square miles of land in northern Quebec, would eventually generate more electricity than twelve large nuclear stations. That would include plenty for export to the USA, from which most of the financing for the $30 billion project had come in the first place. The PASNY 765,000-volt line was to set a crucial precedent for future links to this power.

To reduce the impending threat of a long hearing which might ultimately show their line to be dangerous, PASNY began to put intense pressure on PSC in late 1975 for a quick certificate to build. PASNY purchased over $50 million worth of Italian and other foreign steel and equipment for the line, and then sent secret memos to the commissioners demanding expedited and favorable consideration.

The PSC staff hotly challenged PASNY'S tactics, and PASNY in turn attacked them in the press as obstructionists. Public debate quickly became acrimonious, and, just before the hearings began, PSC received word that they were to approve PASNY's line, quickly, or else. The "or else" was a legislative override of PSC itself. A bill was sent over to the New York legislature which said that the legislature had considered all aspects of this situation and had concluded that PASNY's line would not harm the environment. Therefore, it *ordered* PSC to issue a certificate allowing the line to be built. The bill was quite brief, but its effect would be to undercut the entire regulatory role of PSC. Carey had the votes to pass that bill and the PSC commissioners knew it. The New York Senate was heavily Republican and they were in favor of *anything* that was argued to be good for business. The New York Assembly was heavily Democratic and was controlled by New York City, which is where most of the Democrats in New York are to be found. PASNY's powerline was also highly popular with them because a substantial amount of the power to be carried by the new line would be delivered to New York City, where it would help ameliorate the high rates paid by city residents.

The PSC willow bent in the legislative wind and approved PASNY's powerline. In February they authorized construction, reserving basically only the right to make some design changes that might become necessary as a result of the health and safety hearing.

That was my introduction to the politics of power. No noneconomic force on earth was going to stop or materially alter that line, and that was that. What bothered me most about the process was not that the line itself would be built with no significant protection for the public. If the government, as the representative of the public, had considered the economic factors on one hand and the public health factors on the other and had decided that the economic factors were more important, then the system would be operating the way it's described in the civics texts, and it would be up to the people to change the government if the decision were unacceptable. But Carey and his people made the decision without even *listening* to the countervailing evidence, and then acted as if they had. New York's lieutenant governor, Mary Ann Krupsak, was apparently appalled by the whole episode and split with Carey publicly over it—to the detriment of her political career.

After February 1976 the legal issues that remained were whether the RG&E and NiMo line should be built, and what changes should be incorporated into PASNY's line as a result of health-and-safety considerations. PSC still retained complete flexibility in dealing with RG&E and NiMo, but it was clear that whatever they would ultimately order PASNY to do, it had to be consistent with the line that was already under construction. Thus, PSC's options vis-à-vis PASNY were exceedingly limited.

*

So NAS was gearing up to do a number on Sanguine, and PASNY, with its gubernatorial clout, was shoving it to PSC. Both these developments were to hurt Becker and me in the future.

34

But my immediate problem in February 1976 was the power companies. Their lawyers met in the NiMo headquarters in Syracuse to plan their strategy, and there it was decided that their approach would be to attack me personally...to show that I was a poor scientist and hence that my testimony that thirty other groups of investigators had also found NIEMR-induced effects should not be believed. The only dissent to this strategy came from NiMo, which wanted to concentrate on the evidence itself, and show that none of my reports was really applicable to this particular powerline. The NiMo lawyers actually worked full-time for the company and I think they had a kind of long-term view of what was good for the company. They doubtless realized that even if the strategy advocated by RG&E and PASNY was successful and I was done in, the reports would still be there and they might some day come into evidence through some other scientist; it might even happen in the very same hearing. But Robert Harvey and his RG&E team were outside counsel, and the same was true of the head of the team that represented PASNY, Francis X. Wallace, a law-school professor. I think that Harvey and Wallace cared about nothing but winning *this* case.

Soon after this meeting things heated up. Harvey sent Simpson an incredibly detailed list of demands for information about my research. Harvey wanted every bit of data regarding every experiment I had done since 1974. For each rat and mouse, he wanted to know when and where it was purchased, and its weight when it arrived and every day thereafter until it died or was sacrificed. He wanted to know how each animal died or was sacrificed, why, and the disposition of its tissues. He wanted to know how much each animal ate and drank each day it was in the laboratory, and he wanted every measurement of every animal in each experiment, as well as a copy of each calculation in which any piece of data was used. He also wanted photos of my equipment, and schematic diagrams of the electrical apparatus. Furthermore, Harvey wanted a copy of each report, analysis, or summary about the data written by anyone in the laboratory, and copies of all correspondence we had had with any other scientist about their NIEMR research or about ours. He wanted lists of phone conversations and copies of all memos of conversations. Finally, he wanted his experts to visit our lab to inspect and photograph our experimental set-up.

These demands were, for the most part, an enormous breach of propriety and fairness. For one thing, much of what he was asking for had nothing to do with the hearing; it was my personal business. For another thing, my notes and data contained preliminary observations, partially thought-out ideas, and so forth, which weren't ready to be released. Third, and what concerned me the most, was that my data—like that of most scientists, I would suppose—were taken in my own personal code with my own definitions of various symbols. For example, W_0 meant initial weight, D meant weight change, and C was the concentration of a particular blood-plasma steroid in micrograms per milliliter. Frequently, the same symbol would have a different meaning at different times. The release of my data without a line-by-line explanation of what they meant would simply lead to massive confusion. On the other hand it would take months to provide such an explanation, and that was quite out of the question. Besides, the whole effort would have been stupid because plowing through my raw data would ultimately lead to the core data concerning NIEMR-induced effects in rats and mice that I had already published and given to the power companies. It couldn't be that the companies were merely trying to check up on me, to see that I had actually done the experiments, because if I had fabricated the data that I published, I certainly could have fabricated the raw data to support it. I could only conclude that Harvey's requests were a nasty form of harassment.

I didn't want to appear to be a prima donna, but I didn't like the idea of being pushed around by Harvey—that would have been a very bad way to begin our direct relationship since he was going to be my main cross-examiner. I looked it up, and couldn't find a single instance in which an expert witness was made to hand over his raw data and personal correspondence. Simpson wasn't much help; he was afraid that if I didn't do it I would appear to be an obstructionist and perhaps Matias would throw me out of the case—which, of course, would have made Harvey ecstatic. Simpson's boss, Rheingold, was pressuring him to pressure me to give in. In the end, I did and I regretted it for the rest of the hearing. I learned the hard way that when you deal with somebody like Harvey you never, *never* give in.

At about the same time I agreed to give Harvey xeroxes of everything I had and allow his experts to come and inspect my lab, he hit me with his next demand. He said that the one day's cross-examination to which I had initially agreed would not be sufficient. Simpson asked how much time he wanted, and Harvey told him two to three *months.*

Now Harvey knew that he wasn't going to get two to three months, but I knew he was going to get more than one day. That's the way the game was played. The power companies were always on the offensive, and the PSC staff was usually on the defensive—it was the conciliator and the mediator between the companies and the commission itself. The power companies had no adversary in the hearing in the normal legal sense—someone fighting them tooth and nail on every point—unless homeowners elected to participate, which of course they didn't because they couldn't afford to. So, if the status quo on any given point was unacceptable to the power companies, their tack was to make an outrageously excessive demand; and the PSC staff then worked out a "compromise' somewhere between the status quo and the new demand.

I told Simpson two days...and that's it. That was already more than the time allotted for cross-examination in most PSC cases, and these other experts had been *paid* for their grief. When the issue came up, I was going to allow two days, and then beg off by saying that my duties at the VA didn't permit me to spend more time on the case. That would put the burden on Matias, and he could not very well judge my position to be unreasonable and hence throw out my testimony. Even if he *did* toss me out, deep down I think I might have welcomed that decision; it would have been a face-saving way out of the hearing, with me intact, unscathed, and ready to fight another day in a forum where my chances were better.

In March Harvey sent out his witness team to inspect our lab. Their purpose was to try and find things to zing me with on cross-examination, and I was certainly not going to help them, so I arranged things so that I was between experiments when they came. There were no animals in the room, and the equipment was shut down.

Six people came, including Michaelson and Miller, but not Schwan. "Before we go into the lab," I said, "I need to know if there are any lawyers here." "Nope, Andy," Miller said, "none except you." My being a lawyer bothered him, and that made me feel good.

Miller was the focus of my attention that day because, unlike Michaelson whom I believed I knew very well, Miller was an unknown quantity. He was about forty-five and good-looking, and had what struck me as a kind of insincere sincerity about my experiment. He was a botanist and had never worked with animals; from the questions he asked I figured he'd never even been in an animal research facility. Miller asked me to explain how the electrical apparatus worked, and I did. He nodded as I described it but I was certain he didn't understand what I was saying; when he spoke, he jumbled all the electrical terms I had used just as a beginning student might do. I was tempted to ask him about his testimony but I didn't because I feared it might destroy the mild rapport we had established. So, for the balance of the visit, I continued to try to appear as if I took his scientific position seriously, and he continued to ask softball questions and take a lot of pictures.

*

As the start of cross-examination in April approached, two things happened that raised my spirits a little. First, Michaelson essentially short-circuited himself as a credible witness in the hearings. He had been holding himself out as an expert in every area of concern including the effect of powerline NIEMR on cardiac pacemakers, and on the consequences of failure to properly ground fences and farm equipment near the powerlines. Thus he had already testified several times during the Step 1 and early Step 2 hearings and had fared very badly during the cross-examination by Simpson. Michaelson had a tendency to answer questions in half sentences, and he would refuse to concede the cross-examiner a point, even when the cross-examiner deserved it, if Michaelson figured the admission might hurt his client. It became increasingly clear that he would not seriously affect the outcome of the hearing. This circumstance reduced the power company witnesses with whom I had to be concerned to Schwan and Miller.

Second, just as the cross-examination was scheduled to begin, I got a very interesting phone call. Identifying himself as a researcher in the medical school at Temple University, the caller, Joseph Noval, said he had heard of our experiments with rats and that he had done very similar work for the Navy in the early 1970s—at Johnsville. He had replaced Kendrick as the Principal Investigator of the project. Kendrick, and then Noval, had conducted studies over a period of years that turned up practically the same results as my own experiments on rats. In connection with the Sanguine program, Noval's group had exposed rats to NIEMR far weaker than that of powerlines, and had found stunted growth and altered levels of brain enzymes. Our two groups had done nearly identical research projects, except that Noval had used much weaker NIEMR levels, corresponding to a strength at about 2000 feet from the 765,000-volt line. Noval told me that he had kept a copy of the data and that he would send it along.

Despite Michaelson's self-destruction and Noval's gift—which arrived in late March and turned out to be as good as he said it was—I was still up to my neck in problems. The hearing had become my occupation and preoccupation; it had pushed my research into the background. The VA officials had agreed to allow me to participate as a kind of public service, but they probably had no idea of the amount of time that I had been required to devote to it.

37

CHAPTER 8

In the Dock: Them

With Michaelson defused, the heart of the cross-examination phase of the hearing would consist of Miller and then Schwan for the power companies, followed by me and Becker for the PSC staff. Three days before the hearing began Simpson drove to Syracuse and we worked at my house and my lab preparing for the big event.

Miller's pre-filed testimony that the 765,000-volt lines "did not pose an unreasonable risk to health" rested principally on his analysis of the Sanguine studies. But there were gaping holes in what Miller had written. The Sanguine studies couldn't logically be used to defend high-voltage lines because Sanguine fields were many thousands of times weaker than powerline fields. Besides that, most of the Sanguine studies had actually found biological effects, and that obviously was not good for the power companies. Miller had pointed to a series of negative experiments, but one can't prove very much from negative studies because they amount to no more than a dry hole—they say only that there is no oil at that particular location, not that oil doesn't exist. And even the negative studies Miller cited had actually been repudiated by the Navy itself because they were so poorly done. If all that were not enough, there was the fact that Miller was hardly an expert. He was a botanist, with no training in medicine, biophysics, or electrical engineering.

On April 15 Miller took the stand. Apparently rankled at having to travel to Syracuse, Matias did not even show up. The judge was Harold Colbeth.

Simpson began his questioning of Miller by concentrating on his credentials. Apparently feeling that Simpson was merely trying to embarrass him, Miller reacted testily. By the time Simpson began getting to the issues, Miller was very much on the defensive.

Simpson asked him what he meant by "unreasonable risk."

Simpson: When you say the [NIEMR] doesn't pose an unreasonable risk to public health, you mean that there exists some risk but that medically speaking there is no danger to public health?

Miller: No. I don't mean that at all...to the best of my knowledge, there are no effects from these lines in terms of biological effects...to the best of my ability, I have not been able to determine that there are effects underneath these lines...therefore, I say in the absence of any demonstrable evidence at this point, it seems to me there is no unreasonable risk to public health.

Miller's notion of "unreasonable risk" was clearly tied to the absence of observed effects along rights of way—what I called the Operating Experience argument. Since there had been no studies of exposed human or animal populations near high-voltage lines, the absence of observed effects was hardly surprising. As far as "medical" risk was concerned, Miller was of course quite unable to make judgments on such a matter. (To our surprise, no one picked up on our use of the phrase "medically speaking," and we went right on using it all day.)

As the day wore on some unanticipated practical problems arose. For one thing, the power companies had five lawyers and one of them was always objecting to something. Although I sat at Simpson's side and frequently conferred with him during the cross-examination, I was not permitted to speak to anyone else. Thus, it was Simpson alone who had to respond to the various power-company lawyers and to the judge while trying to maintain the thread of his cross-examination. His efforts, although valiant, were not always successful.

For another thing, Colbeth was very weak. He had been the loser in a recent power struggle within PSC and had been demoted from a division leader to a judge. Because of state civil service rules he had kept his previous salary of $42,000 a year, which was about $2000 more than Matias, who was an experienced judge, had been earning. I think Matias resented it and the relationship between the two was not good. Furthermore, Matias had actually established the rules for the hearing during 1975; Colbeth had been brought in well after things were under way. Finally, Colbeth wasn't a lawyer, and he seemed to know little about the scientific issues. Thus, he was unable to establish real control in the courtroom.

This vacuum tended to be filled by the power-company lawyers themselves, or by the witnesses, and it resulted in some unfortunate legal situations. The worst arose from a precedent which Miller set. We had asked him a specific question about one of the experiments to which he had referred; in response, for forty-five minutes, he read a prepared attack on the experiments of Dietrich Beischer. Simpson could not turn off the spigot. This sort of thing eventually led to an absurdly long and often confused hearing record.

The third problem was Miller himself. He wasn't very bright, but he was tenacious as hell. He reminded me of one of those lizards that bites and then holds on mindlessly even when you pound on its head with a bat. He fought Simpson on practically every question—more like a lawyer practicing adversarial science than a scientist—and in his refusal to concede what he thought would hurt Harvey's case, even when the facts demanded a concession, he was a carbon copy of Michaelson.

All these factors suggested that we were in for a long haul with Miller.

The beginning of the second day confirmed that. It started out with Simpson asking Miller whether he knew that the Sanguine fields were 100,000 times weaker than those under the powerline. After Miller agreed that "that seems reasonable," Simpson started trying to steer him back to his own conclusion that the negative Sanguine studies showed transmission lines were safe. Miller seemed to sense what was coming, and the power-company lawyers, particularly Harvey, began to object to Simpson's questions, demanding repeatedly that he clarify the question, or withdraw it, or rephrase it, or go on to a new topic. It took an hour of contention for Miller finally to concede the point: the Sanguine studies, he said, "have reduced relevancy in terms of not answering the question" at high NIEMR levels.

Simpson then began to attack the negative Sanguine experiments on which Miller actually relied, a series of nine studies performed at Hazelton Laboratories under contract to the Navy during 1968–1970. Miller had described them approvingly in his testimony and had made them seem like full-fledged experiments, when they were in fact self-described "pilot studies."

Moreover, Hazelton Laboratories was under Federal investigation for rigging research for drug companies. Several years before Miller's testimony the Navy itself had stopped citing their studies in support of its position that Sanguine would be safe. In view of this situation, Simpson wanted to make Miller admit that his testimony was "largely dependent" on the Hazelton studies. Harvey fought tooth and nail to inhibit this line of questioning, and in the process he turned the courtroom into a circus.

Harvey: Mr. Simpson, by "largely" are you referring to the number of words or the number of pages?

Simpson: Do you understand the question, Dr. Miller?

Harvey: Mr. Examiner, if Mr. Simpson isn't going to qualify that, I will object to the question as not being clear.

Simpson: I think the question is very clear, and it can be answered. It is very qualitative. I think Dr. Miller will be able to answer the question.

Harvey: I do object to that, Mr. Examiner, because I think that the problem here is whether we're evaluating the significance of the critique or the number of words devoted to the Hazelton studies, and I think that it's very necessary to make sure that we know whether or not Dr. Miller is to respond to whether a majority or large number of pages are devoted to the studies or whether he believes that those pages in terms of qualitative analysis are largely devoted to that.

This kind of doubletalk issued from Harvey whenever he sensed that his witness was in trouble. In the beginning Simpson actually responded to the substance of Harvey's objection. But then Harvey would object that Simpson was mischaracterizing his objection and demand that Colbeth order the court reporter to read the question back. At this point, the hearing would stop as the reporter searched back through the transcript, found it, and read back Harvey's objection. Then both Simpson and Harvey would claim that the re-read transcript supported their respective positions regarding the merits of the objection. This nonsense could easily continue for thirty minutes, and the whole time Colbeth would just sit there, doing nothing. At the end Miller would be rested, collected, and ready to go, and Simpson would have a blood pressure somewhere off scale.

The only way out of this dilemma was to stop arguing with Harvey, and that's what Simpson began to do. Miller finally had to concede the point: "It looks like the majority of words [in his testimony] are pertinent to the Hazelton studies." But it was only because Simpson wouldn't let go. He impressed me a lot. He need not have worked so hard.

Now that Miller had admitted his testimony was "largely dependent" on the Hazelton studies, Simpson wanted him to concede that the studies were poorly done. We were treated to the same dog-and-pony show, but in the end Miller gave us what we wanted: "The experiments were not, I would say, in my opinion, at a professional level." He even went on to criticize some of the individual experiments: they were "faulty," "not well designed," or of "very poor design." These were experiments that *he* had cited in his testimony; thus Miller knocked away his own support for concluding that the 765,000-volt lines would be safe.

By the end of the second day, there was a general feeling that we had made our point with Miller and that it was time to move on. But Simpson and I had another series of questions in mind, so Miller was back on the stand on Wednesday. Our basic idea was to bring up several of the Sanguine studies that he had *neglected* to mention in his testimony, and to force him to agree that they suggested that the 765,000-volt line might be a health risk. Simpson and I didn't get that, but we did catch Miller in a vise. What was ironic was that it involved experiments with slime mold.

Slime mold, a primitive but very interesting life form, had been exposed to Sanguine NIEMR in a series of experiments by Dr. Eugene Goodman and colleagues at the University of Wisconsin. These tests resulted in several changes in growth and reproductive characteristics. When we raised these studies with Miller, he testified that Herman Schwan had told him the NIEMR strength used by Goodman was so high that the experiments applied neither to Sanguine nor to high-voltage powerlines. Miller lavished praise on Goodman's work, because (as was apparent to me) he could afford to do so and still not hurt his client's case. But Schwan was a biophysicist and Miller was a botanist, and there was a huge intellectual gap between them, particularly when it came to calculations. As I sat there listening to Miller drone on and on about Goodman and about Schwan's calculations, I thought of a way to exploit their differences. During a break in the hearing, Simpson and I worked out a series of questions to show that if Miller had *properly* understood Schwan, the calculation would show that Goodman's NIEMR level was *below* that produced by powerlines.

As Simpson pursued this line, sure enough Miller began to soften. First he said that there was a "little problem" with the way the experiment was conducted, and that it had a "lack of appropriate controls." Then he said the results "may well be an artifact." Eventually he said: "Now I am criticizing the experiment, saying it was not a properly controlled type of experiment."

I realized Miller had contradicted himself, so as Simpson stalled for time, I frantically paged through my notes to find Miller's exact words concerning Goodman's work—spoken when he believed the study did not apply to powerlines. I found them and gave them to Simpson, and he did his thing.

Simpson: Dr. Miller, isn't it correct that you previously stated "For example...the work of Dr. Goodman. I think this is an outstanding study... here is a beautiful example of a well-controlled, well-analyzed, a beautiful experiment." Did you not so state?

Miller: Yes, I think that's correct.

41

That was the only time the courtroom was silent for more than a moment with Miller on the stand. Colbeth then called a recess, and Miller got up slowly and walked down our side of the room. He stopped behind me and Simpson and said, "I guess you guys got me there." "Looks that way, Mort," I said calmly.

That night Simpson and I celebrated, and my grandmother cooked us a home-made spaghetti dinner. But I kept wondering about the Goodman episode and about Miller. How many scientists had the power companies had to contact before they found him? How much were they paying him to do what he did? (I had discovered during the flap over the NAS committee that he owned over $10,000 worth of stock in Niagara Mohawk, the company for which he was testifying.) He was simply unbelievable; however, his performance on the stand was an instructive episode in my suddenly accelerated education about the politics of science. I remembered in particular another comment he had made about the Goodman study, a hysterical kind of attempt at humor. He had theorized that because the NIEMR caused the cells to divide more slowly, it might mean that NIEMR would *lengthen human life spans*. Unable to stop himself, he had rambled on, "If that's true, let's string them all over the God—the countryside!"

Miller was actually on the stand for two more days, responding to questions about other Sanguine studies that he had neglected to mention in his testimony, and admitting that one study, commissioned by a power company, had actually been rigged. His lawyers constantly bitched at Simpson, and they succeeded in slowing us down to a crawl. I realize now that it didn't matter in the least, because Miller was dead, and it made little sense to continue thumping on him. But we had tasted our first blood, and we wanted more.

*

Following the abject Miller, the imperious Herman Schwan took the stand. It was like the difference between beauty and the beast. Looking at Simpson and me, Schwan probably figured he had nothing to worry about; we were the rookies, and he was the master, giving his trusted and true testimony. It was the same logic and analysis that he had given to the highest levels of American government for twenty-five years—why *should* he be worried?

Schwan had said in his testimony that since calculations showed that powerline NIEMR wouldn't cause heating, and since there were no other known effects of such NIEMR, the 765,000-volt line would be safe. His major premise was a complete red herring: nobody in the entire hearing cared about heat effects because we all knew it was essentially impossible for powerlines to cause heating. So our plan was to attack Schwan's minor premise and his conclusion.

The minor premise in Schwan's syllogism was that there were no known nonthermal NIEMR-induced biological effects. But I had found thirty-two reports of such effects, so clearly either Schwan was wrong, or "40 million Frenchmen" were wrong.

After asking Schwan to sketch out his general criteria for validity in scientific experiments (they were exceptionally rigid), Simpson moved to the studies at hand. We picked mine first because we thought that with me sitting there, Schwan might be more subdued in his criticism.

Simpson: Dr. Schwan, does Dr. Marino's study meet the criteria you listed previously...?

Schwan: Not necessarily...there is always a possibility when you conduct an experiment such as Dr. Marino did, that it can be explained in two ways. One way would be that there is a subtle [NIEMR] effect acting in an unknown fashion directly, and there is another possibility that artifacts are involved.

Schwan went on to say that he thought the changes in body weight and body chemistry that I had measured in the rats might have been due to tiny shocks that the animals received, not to the NIEMR itself.

I myself had written that it was possible that the effects were due to shocks, though when I evaluated the totality of the evidence and data, I concluded that the effects were probably due to the NIEMR. I had explained in both my paper and in my testimony how I had arrived at that conclusion. I wondered if Schwan had actually read my paper on the experiments.

Simpson moved on to the work of Rutger Wever, a German researcher who had shown that very weak NIEMR levels could cause disturbances in human biorhythms. His volunteer subjects had lived in an underground bunker for up to eight weeks; during that time Wever measured the daily rhythmic fluctuations in such things as body temperatures and blood levels, and found that these were significantly altered. Simpson asked Schwan if Wever's studies measured up.

Schwan: ...[It] disturbs me somewhat...that Dr. Wever has conducted more than 100 experiments. Only a small fraction of these experiments, something like 20 or 40, I forget the precise number, have been published. Why not all? Is it perhaps a subconscious desire of selectivity?... It may be or it may not be. I find that only a fraction of his experimental results are reported.

Simpson: Is it your testimony that he is holding back valid results?

Schwan: I'm not saying he's holding back....

Wever of course was not present to defend himself. It was *possible*, I suppose, that Wever had fudged his data. I only knew him through his published scientific studies, but Schwan knew him personally. Although I didn't believe Schwan, Simpson and I were in no position to call him a liar because we couldn't prove that Wever hadn't hid his data. So we moved on.

Working at the University of Maine, James McCleave had studied fish perception, behavior, and physiology for many years. His eel and salmon experiments were classical and elegant, employing a technique long used by researchers in the field. First, the fish were immobilized in a tank so that McCleave could continuously record their heartbeats. Then he applied an extremely weak NIEMR signal to the fish, and several seconds later a small electric shock. After many such trials, the fish began to associate the NIEMR with the subsequent shock, and this perception was manifested in an increase in the fish's heartbeat. McCleave conclusively established that the fish could perceive the NIEMR itself because its heartbeat increased with the application of the NIEMR even after the shock had been discontinued. Simpson asked Schwan whether the study satisfied his criteria, and Schwan replied, "Not necessarily." He said he felt certain that with the fish swimming around in the tank, some artifacts must have been created. But McCleave's fish had been prevented from swimming during the measurement precisely to avoid artifacts.

Simpson: Is it your testimony that the fish were swimming throughout the period of measurement?
Schwan: I have to look that up, I cannot comment on that. May I check on that?
Simpson: Yes, go ahead. Check it.
Schwan: Mr. Simpson, I cannot find right away any information which pertains. I would need more time.

Simpson went on to another study, in which Rochelle Gavalas-Medici had found that the reaction time of monkeys had been affected by very weak NIEMR. Simpson asked, "Doctor, does this study meet the criteria you have listed?" Again Schwan replied, "Not necessarily." He said he couldn't understand Gavalas-Medici's report because it wasn't clear and that she was "incompetent" to clarify it.

Next Simpson asked Schwan about a bird study done by William Southern. Southern had built a large cage on the ground directly over the buried Sanguine test antenna in Wisconsin. In the center of it he placed young gull chicks, and he noted the direction in which they moved, knowing that they have an instinct to move more often in one particular compass direction. With the antenna turned off, after thousands of individual trials, Southern found the compass heading for the birds. Then he turned the antenna on, thereby creating a NIEMR environment, and found that the birds failed to show a preferred heading and dispersed randomly in every direction. Southern showed by standard statistical techniques that the difference in behavior was significant, and he attributed the effect to the NIEMR.

Simpson: Doctor, does this study meet the criteria?
Schwan: Yes and no. Where it is lacking is in the statistics…it is a lack in the significance of the numbers. In other words, I personally could not see a really significant difference between the controls and the experimentals.

But didn't Southern conclude, nevertheless, that the NIEMR was sufficient to disrupt the birds' orientation? "Not necessarily," said Schwan.
And that's the way it went all day long. With study after study Schwan found some possibility of error or irregularity. Schwan's minor premise wouldn't have held up if he had accepted even *one* study as valid, and he knew that.
As far as the hearing was concerned, it was now a matter of credibility. Schwan had judged the work of investigators in many disciplines far outside his own—physiology, biochemistry, psychology, genetics, ornithology,

44

ichthyology, to name a few—and found it *all* lacking in quality, rigor, and in some cases honesty. I thought one would have to be an idiot to believe him.

On the second day we went after Schwan's conclusion. We told him that we would assume, for the sake of argument, that there were no biological studies that reported effects due to NIEMR. We also told him that we would assume his calculation was correct and that, therefore, people would have NIEMR levels inside their bodies of 0.1 milliamp/square centimeter (mA/cm2) from being exposed to the NIEMR of the powerlines. Finally, we would assume that 0.1 mA/cm2 would not cause a person's body tissues to heat up. Simpson and I wanted to make the point that there were countless processes that went on in the human body with respect to which *no one* had ever studied the consequences of imposing NIEMR levels of 0.1 mA/cm2, so that it was impossible to say what the consequences of that NIEMR level might be.

Simpson: Now, if it were true that 0.1 mA/cm2 could not cause heating (or shock), would that alone mean that such a current density was safe?

Schwan: Within the framework of my knowledge, I'd have to say yes.

Simpson: Would knowledge of the [NIEMR] level in bone permit you to predict whether or not bone tumors will occur?

Schwan: It certainly will not permit me to do so.

Simpson: Will knowledge of the [NIEMR] level in bone permit you to determine whether the modeling rate of bone will be affected?

Schwan: No.

Simpson: Doctor, will knowledge of the [NIEMR] level in bone permit you to predict whether fracture-healing would be affected?

Schwan: No, it won't.

Simpson: Doctor, do you know the mechanism underlying the modeling of bone or fracture-healing in bone or production of bone tumors?

Schwan: No, no one does.

Simpson: Doctor, if the mechanism for these functions is not understood, how could anyone conclude from the knowledge of the [NIEMR] level in those tissues that there will be no effect on those functions?

Schwan seemed not to know how to deal with that. He told Simpson he did not understand the question. So Simpson rephrased it a bit more directly.

Simpson: Now, Doctor, if knowledge of the mechanism for growth control of bones, remodeling rate, and bone tumors is unknown, how do you know that the current density of 0.1 mA/cm2 will be safe with respect to these biological processes?

Schwan's usual icy reserve began to give way a little. He tried to ignore Simpson, retorting to Matias,

Schwan: I submit that the question which Mr. Simpson formulates is nonsensible, and the reason why I say so is that without any qualification he talks about bone growth and relates it to current density...he talks about two entirely unrelated phenomena...utterly nonsensical.

Schwan tried to fight off this attack by becoming the questioner himself.

45

Schwan: Mr. Simpson, how does bone growth, normal bone growth for example, relate to current densities? Can you explain that?

Simpson: Your Honor, this is the expert that is saying that 0.1 mA/cm2 is safe. Now, I am asking him to tell me how he knows that it is safe with respect to bone growth, and I think I am entitled to an answer to that question. I did not make that statement. The witness has.

Schwan continued to react as though Simpson had no business asking the question, but before he could go very far, his lawyers requested permission to speak to him off the record. After a brief conference, they returned to their seats and the questioning continued.

Simpson: Doctor, if you do not know the mechanisms for bone growth and bone tumors, how do you know that a current density of 0.1 mA/cm2 is safe?

Schwan: I still have trouble to understand your question. For example, you might come with the following question to me: If you do not know when the sun will set (sic) this morning, how can you know that current density is safe? Can you kindly explain to me what bone growth has to do with current density?

Simpson: Doctor, are you asserting that there is no relationship between current density and the process of bone growth?

Schwan was in trouble again because Becker and several other research groups headed by orthopedic surgeons had each successfully employed weak NIEMR to grow bone in patients and hence successfully treat some orthopedic diseases.

Throughout Schwan's difficulty, the tension had been mounting in the courtroom and now again it erupted into an argument among the lawyers. The strain on Simpson was very great. He had to contend with up to four lawyers, confer with me after nearly every answer from Schwan, and maintain a discipline in the flow of his questions. Then, Simpson asked the question for what seemed like the hundredth time.

Simpson: Assuming a current density of 0.1 mA/cm2 in the bone, would those levels be harmful with respect to bone growth, bone tumors, or bone remodeling?

Schwan: I don't know and I can say why I don't know. The literature available to me, and I refer to the publications of Becker, Bassett, Brighton, and others, does not provide sufficient information to answer your question.

That was it...that's what we needed and wanted...he *couldn't* say his "safe" level was really safe because he didn't know whether it would affect the tissue. So the next thing was to expand on our beachhead. I had prepared a list of fifty processes that involved virtually every organ in the body. Simpson pursued the same line of questioning involving the lung, the kidney, the brain, and the pituitary, thyroid, parathyroid, adrenal, and pineal glands—all leading to answers that said, in effect, no, Schwan didn't know. Finally, Schwan had had enough. He told Simpson "That's it," slammed his books closed, and folded his arms. "No more questions, your Honor," Simpson said.

I saw the thirty-two studies in my testimony as a kind of developing flower that needed to be nurtured, protected, and cultivated. These studies pointed to a new biology, beyond the bounds of what we knew at present. Any shortcomings in them had to be viewed compassionately and sympathetically because they contained, possibly, the seed of something new. Nothing in science, particularly biology, ever emerges full blown and correct in every detail. But Schwan seemed to view the studies as weeds in the garden of known physical laws. Because the known laws did not predict the results of the studies, Schwan's reaction was to denigrate them, rather than assume that there existed unknown laws, or unknown interpretations of known laws. Neither one of us had much respect for the approach of the other. We each thought the other was unscientific.

CHAPTER 9

In the Dock: Us

Now it was my turn on the hotseat. Simpson and I had made a mistake with Miller because we kept him on the stand well beyond the time we needed to. That blew my position that two days for me on the stand was enough. We had had Miller and Schwan for a total of seven days—that was Simpson's argument. If I chose too low a figure, the power companies might argue successfully that they were being denied their right to cross-examine me, and Matias might have thrown me out of the case. I wanted to give three, then five, then even six days, but Simpson felt that it was not enough.

Finally I decided on eight days. That was longer than it took for the creation of the world, and certainly enough time for Harvey to try and break me. He had received a written copy of my testimony almost five months earlier, and he had virtually unlimited resources to help him in picking it apart. I thought eight days was obscenely long, and indeed it was longer than any witness had ever been cross-examined in the history of the New York PSC.

I showed up at the hearing on the first day as ready, I thought, as I would ever be. I brought each of the studies or reports I'd cited as reference in my testimony, as well as every document or report that I thought I was likely to need. Simpson had tried to learn from Harvey, as a courtesy to me, which areas of my testimony would be covered on which day; that would have permitted me to reduce the amount of material I needed to lug to the courtroom each day. But Harvey refused to tell Simpson anything, and I had to bring each page of information to the courtroom each day. It was a total of six large cardboard boxes, and it took two trips with a handtruck to move the material from my car to the seventh-floor hearing room. The power-company lawyers and their advisors each left their material and papers in the courtroom at night because it was locked. But my blood was in those boxes and I didn't trust anybody on the other side, so I lugged them in each morning and out each night. I never let them out of my sight; even when I went to the men's room, someone stood guard over them.

The courtroom was a large conference room in a state office building in Syracuse. At the right was a long line of tables for the power company people. They had seven lawyers on the first day of the hearing; sometimes they had more, but never fewer than four. Miller and Michaelson were there to help with the analysis of my answers the same way I had helped Simpson. The power companies also had three electrical engineers, several gofers, and some other people whose role I couldn't determine. On the left side of the room sat Simpson and one PSC staff engineer.

Almost no press people or members of the public were present. That was hard on me because I thought the general public was my natural constituency.

48

I was there, I felt, representing them, trying to help see that they didn't get screwed. But this was Syracuse, and it was a work day, and the people who would be most affected by the line lived hundreds of miles away. The Syracuse newspapers editorialized in favor of the 765,000-volt lines because of what they said would be an economic benefit to the upstate New York area. And usually the news items in these newspapers were prepared from power-company press releases. These releases were typically after-the-fact summaries of what the power companies saw as the important facts and the meaning of the event. Given all this, I was psychologically very much on the defensive.

In the beginning I was afraid, a vague kind of a feeling that you get when you're doing something you consider important but that you've never done before. As the process wore on, I became fatigued and my principal emotion was resentment.

*

Harvey was a clever and persistent questioner. In the beginning he had the advantage because he had cross-examined witnesses before and I had never been cross-examined. As a result I made several early mistakes in responding to Harvey's questions.

One mistake, a bad one, was to give an exhaustive answer to a simple question. Harvey asked me, for example, what my conclusion of a health risk from exposure to the NIEMR of the powerlines was based on, and I told him the thirty-two studies in my testimony. He then asked me to explain the studies. Like a fool, I started to do that; I started to explain each of them in detail. That involved the description of the experimental setup, the number of animals used, what was measured, and so on. I frequently had to stop and spell the name of an author or some scientific term so that the court reporter could get it straight. I had talked continuously for about an hour and a half, covering only about three studies, when I realized suddenly that my batteries were running down. I was getting hoarse and tired, and Harvey was just sitting there, with a Cheshire-cat grin on his face.

What I was saying couldn't possibly hurt Harvey's case because the information was already in the record in my testimony. What was important was not *what* I was saying, but the consequences of the process of my saying it. When I realized this, I told Matias I had to go to the men's room and he called a recess in the hearing. I collected my thoughts, and when the hearing resumed I told Harvey that I had thought about it during the break and had realized that the answer to the question was already in my written testimony, so that it wasn't necessary for me to repeat it on the stand.

Harvey raised hell, claiming he had a right to hear the answer from my lips, saying I was being evasive, and that his client's constitutional right to cross-examine me was being denied. He said that it was absolutely essential that he have the added detail that I could provide by verbally describing the studies. He argued the point interminably with Simpson, but eventually Harvey lost.

Another thing he did was ask an open-ended question, like asking me to explain "the scientific method." It might take me ten minutes to respond, and in the process I would make a number of factual statements to help explain my answer. As soon as I made one of those statements, it was "testimony," and even if it had nothing to do with powerlines Harvey would ask me questions about *that* statement to try to show I had made an error. The problem, I gradually realized, was that I was treating the cross-examination as a bona fide inquiry into the truth of my testimony. But Harvey was primarily interested in making me look like a nincompoop, and one way to do that was to show that I made errors, *regardless* of what they involved. So I began to give very brief, specific answers. I would say, for instance, that the scientific method is a method that scientists have for finding the truth. If Harvey asked me to explain, I'd ask him what *he* wanted to know; if he couldn't verbalize the question, then he wasn't going to get an answer. After that first day, all my answers were narrow and specific. It made Harvey angry and he accused me again of being evasive and argumentative. So? I was learning the rules of the game from him.

I began to conceptualize the hearing as a contest in which Harvey would reach down into a big bucket of dung beside his chair, scoop out a handful and throw it at me. I had a jai-alai basket strapped to my arm, and as the dung flew I caught it and hurled it back at him. My aim was to see that every shot rebounded. But eight days of this game began to look like an eternity.

Another thing that caused me trouble, especially in the beginning, was the equivocal use of words. For example, the word "speculate" is scientifically respectable, and frequently used; one might say, for example, that Einstein speculated on the physical nature of the universe. In law, however, the word points to unreliable opinion; one might say, for instance, that a witness offered no evidence, only speculation. When Harvey used the word in the context of a scientific discussion, I would automatically invoke its scientific meaning. Then in his follow-up questions he would switch to the other meaning and suggest that I didn't *really* know what I was talking about, but that I was only guessing. In some instances he would corner me so that in any answer I gave, I would be automatically invoking Harvey's meaning (the old "When did you stop beating your wife?" business). For instance: "Did you communicate what you have now admitted was only speculation concerning health risks of powerlines to anyone prior to 1974?"

Another such word was "significance." In science, the word has a precise meaning. A result is significant if the odds against it being due to chance are less than 1 in 20; if they are more than 1 in 20, we say the result is not significant. A scientific result can be significant but not important, or it can be important but not statistically significant. I got into trouble when I allowed Harvey to mix up the two meanings; I might know what I meant, but how about the judges?

The same word game was played with "expert," as in, "Are you an expert in statistics?" If I said "no," he would suggest that I was an amateur at evaluating my data. If I answered "yes," he'd establish that I hadn't written any textbooks on statistics and then suggest that I wasn't really telling the truth.

I finally decided never to respond quickly to anything Harvey said. I would wait a few seconds, and play the question back in my head. Was the question true? Did it contain a false statement which I would be seeming to adopt if I gave *any* answer to the question? Whenever this was the case

I would tell Harvey his question wasn't true. That drove him crazy. He complained that he'd never heard of such an objection from a witness; how could a question not be true, if it was just a question? If the question passed that first test, I examined it for words that might have a dual meaning. If I found one I'd ask Harvey what he meant. When I first did that, Harvey would ask me what I meant by the word, and I'd tell him it depended on the circumstances. You used it first, I'd say—*you* define it.

By the second or third day I was pretty much in gear, and for sheer gamesmanship I began to get the upper hand. After I thoroughly analyzed each question for truth content and for tricky words, if there was any doubt whatever about the meaning I would ask the court reporter to read it back. Harvey was verbose and given to long-windedness and complication, often at the expense of logical syntax or grammatical sense. Most of his questions were written out in advance, so that *he* knew what he was saying, but once the words left his mouth they actually belonged to the court reporter, who was frantically trying to take down a verbatim transcript. When the reporter read back the question there was frequently some ambiguity or misunderstanding because of the complexity of the question, and we'd have to go off the record while the lawyers, Matias, and the reporter straightened it out. Then we'd come back on the record and the reporter would read the question using the language Harvey had originally intended. After all that, if I wasn't absolutely sure I knew the answer I'd tell Harvey, "I don't know."

Three or four other lawyers asked me questions from time to time during the eight days, but none of them was in Harvey's class when it came to being sneaky. I went through the same mental processes with them that I did with Harvey in order to protect myself, and I found that they generally lacked Harvey's tenacity. It often seemed they didn't really want to ask the question in the first place, and frequently they would simply withdraw it and go on to something else. Harvey was the only lawyer who, it seemed to me, had actually prepared for the hearing. Most of the others asked orthodox questions that any lawyer could ask of any witness regardless of the subject matter, or questions that would spontaneously suggest themselves for one reason or another.

The lawyer who most fit this latter category and my chief antagonist in the hearing after Harvey was PASNY's chief lawyer, Francis X. Wallace. Wallace was tall and very distinguished looking, and he sounded like a lawyer because he was constantly using *irrelevants, immaterials, subpoenas duces tecum,* and other legal mumbo-jumbo. Matias had an almost fawning respect for Wallace, which I found very strange.

It was usually hard to take Wallace's bluff and bluster seriously. His cross-examination consisted mostly of asking me about my personal correspondence. First was a letter I had sent to a union official in Canada who had written to ask me what I thought about the efforts of the power companies to try to find the truth about health risks and powerlines. In my letter I had told him not much. It turned out that the man was connected with Canadian power companies, and apparently Wallace had obtained a copy of the letter from them. He wanted to show I was prejudiced against power companies, having made up my mind even before the hearing, and therefore I shouldn't be testifying.

Wallace also had a copy of a letter I had received from the engineer who worked with Donald Gann at Johns Hopkins. I had sent him one of the units I had used for exposing rats to NIEMR and he had agreed to make measurements on it and compare them with the values I had calculated. In his letter to me he said that he felt my calculations were wrong, based on his measurements; he sent a copy of the letter to the power Companies because he and Gann were required to do that by their contract with them. Wallace wanted me to admit that my calculations were wrong, and that the NIEMR level of my unit was much higher than what I described in my testimony, and therefore my results didn't apply to powerlines. But it had turned out that it was the engineer's measurements that were in error, not my calculations. Wallace had no copy of *that* letter.

That was basically Wallace's whole cross-examination. I found it hard to take seriously and my disrespect began to show through: I started drawing sketches of various people in the courtroom. That greatly irritated Matias in particular.

Wallace and the other lawyers were basically an intermission to Harvey's Act I and Act II. As the hearing continued I saw a few new tricks from Harvey, but they were crude and transparent. For instance, he would try to bait me with sarcastic comments. After I'd finished an answer he might say something like, "Thank you, Dr. Marino, we all know how interested you are in the plight of the little man," or "Well, I guess there's nothing that you aren't an expert in, is there, Dr. Marino?" The tone and timing of these insults made me want to reach across and rap Harvey in the mouth—or at least reply in kind. But as Simpson kept reminding me, that's exactly what Harvey wanted. So, I learned to smile, and smile, and Simpson's worst fears and Harvey's fondest dream never materialized.

The eight days on the stand were a little legal procedure and a pinch of science, but mostly theater. My best performance came on the seventh day. It was partly planned and partly ad lib; I doubt either Harvey or Miller will forget it. It began in the morning when Harvey asked what seemed like the millionth stupid question about my rat and mice experiments. By then I had received a copy of Joseph Noval's report concerning the Johnsville rat studies. These studies, like mine, were preliminary and tentative, but they each pointed to the same conclusion—that NIEMR could affect the growth rate of rats. I assumed Harvey knew nothing about Noval's report, and my plan was to spring it on him when he gave me an opening.

The opening came when Harvey suggested that since no one else had ever observed stunting of growth of rats due to NIEMR, my very provocative results ought to be confirmed before they could be believed. I told him they *had* been confirmed. Almost reflectively he asked me by whom…and that's when I started. I began in a low, methodical, unemotional tone, telling Harvey that rat studies had been performed under the aegis of the United States Navy at a naval research facility in Pennsylvania; that the experiments had continued for more than three years, and had involved exposing rats to powerline-type NIEMR of very low intensity. The Navy investigators, I told Harvey, had used the same kind of animals I had, even the same strain; they had used almost the same kind of exposure apparatus; and they had exposed the rats for almost the same length of time. I went on to say that they had reached exactly the same conclusion that I had, and that neither group of investigators had known of the existence of the other.

By the time I got to this part of the story I was talking louder and pounding the table for emphasis. Then I stood up, pointed at Miller, and said he must have been given a copy of the Johnsville study because he was on the NAS committee. So, I said, either Miller was hiding evidence from Harvey or Harvey was hiding evidence from the court. By now I was shouting and wagging my anger at Miller, saying the situation was deplorable and it meant *somebody* was hiding evidence.

At precisely that moment the fire bell in the building went off and we were ordered to evacuate the building. That was the only time that I saw Harvey at a loss for words. He kept shrugging his shoulders as if to say he didn't know what I was talking about. Miller said that he had got "three feet" of material from NAS and didn't know whether or not the Johnsville study was in it. The courtroom broke up quickly as everybody headed for the exits. I literally bounded down the seven flights of stairs.

*

Now the heart of my testimony was that more than thirty groups of investigators had found NIEMR-induced biological effects in animals, and hence to put people in a similar electrical environment with no protection whatever would constitute a health risk. The experiments showed that the NIEMR environment was biologically active, not a physiologically innocuous thing like the color of one's socks. I felt that my point would have been made if even *one* of the studies was valid. Since they had all been done by reputable scientists at reputable institutions, I thought it beyond reason that they could all be dismissed as poorly done. I had read those studies very closely; I may even have memorized them. Where there was doubt or ambiguity or misunderstanding, I had called the investigator or gone to visit him. Thus I frequently knew more about the study than was actually in the published version. As a result I expected and *wanted* Harvey to question me about these studies. But neither he nor the other power-company lawyers did that. Almost the entire eight days was spent trying to trick or trap me, or to show that I was evasive or argumentative, or that I had contradicted myself or misrepresented things or committed errors.

My time on the stand ended much as it had begun, with the lawyers arguing among themselves. Throughout the eight days Simpson was a rock; he was never bullied and he never quit. He protected me from the worst of the abuse; I don't think I would have made it without him. We lived that experience together and we became good friends. When I left the stand, I figured I had paid my dues for raising the NIEMR issue and now it was over.

*

The only thing that bothered me was that Becker still had to go on the stand. He had an entirely different attitude and a wholly different way of looking at things from mine. He was a teacher; he presumed that if you were in the room when he spoke, you had come to learn. Well, that was *not* why the power companies would be there.

53

I thought I had been on the hot spot, but it was Becker who made the connection between biological effects seen in the laboratories and probable human medical consequences along the right-of-way of the 765,000-volt lines. And he was the only physician testifying in the hearing. These factors made him very dangerous to the power companies.

Becker knew what I had gone through and he was prepared for an unusual effort of focus and control. But he had an additional problem: he had to care for his patients at the hospital. That meant he had to be at the hospital very early in the morning, testify all day at the hearing, and then return to the hospital in the evening.

From the moment Becker walked into the hearing the power-company lawyers were, I thought, abusive and disrespectful. Their pattern was to ask him a general scientific question, and after he responded, to zing him with suggestions that he was only blowing his own horn. For example, they asked him right off about his theory about the way the body controls its own growth and regeneration. After he answered, the lawyers suggested that the theory was old hat, that it had "been known or a hundred years." "No sir," said Becker, "I hope not, because that is the concept that we feel that we have developed."

The most outrageous thing they did was to make it appear that Becker was merely a puppet, and that I was pulling the strings. Becker had said in his testimony that "The first and most obvious conclusion to draw is that NIEMR does have biological effects, some at extremely low strengths." They asked him if that was his conclusion, or if he had simply lifted it from me—sort of the same way Miller had adopted Schwan's calculations. "This is *my* conclusion," Becker said. His tone of voice was such that everyone knew he thought it was a dumb question, and it annoyed him. When *I* would ask him a question that would elicit such a tone, I knew it was best to go on to another matter. But the power-company lawyers didn't go on to a new matter—precisely, I think, for the same reason that I would have.

They kept putting words in Becker's mouth that on the one hand were extreme, but on the other hand (they would intimate) flowed naturally from his testimony. They suggested, for example, that his testimony was tantamount to recommending that all U.S. powerlines be shut down because of health risks. Becker said no, that was not what he was saying. What he was saying was this: "The most prudent course to follow would be to determine the complete spectrum of biological effects produced by exposure to power-line [NIEMR]... I would, however, not turn off the electricity because of the other social factors that would appertain as a result of such an event...but I do stand on my recommendation that a problem, in my opinion, does exist, and that the problem will not go away and that it should be studied. Now, it is quite possible that the results of this study will indicate that we have produced [NIEMR] pollution that equals or perhaps exceeds the chemical pollution that we have produced. I don't know. In that case, then, certain things would have to be done. To draw an analogy...timetables are set up for the discontinuance of pollution sources, and the same sort of principle I think could be applied here."

Becker was courteous and courtly throughout the day despite the provocations. His sarcasm flared only once, near the end of the day, when the power-company lawyers suggested that we had seen a stress response in the rats not because of NIEMR, but because of their fear of being sacrificed. It was unlikely, Becker said, that the animals knew they were "marching to the guillotine."

Becker went to the hospital at the end of the day, treated his patients, went back to the hospital very early the next day and saw more patients, returned to the courtroom to be in time for the start of the hearing, and sat there all day and endured the same foolishness all over again.

When he showed up at the hearing he had a manila folder with a note pad and a copy of his forty-page testimony in it; that was all. Harvey wanted to know why he had not brought copies of the thirty-two reports with him, and Becker told him, "That's what we have libraries for." There was no way in the world Becker was going to sit there and discuss every detail of each of the reports—they were done by good scientists, at good institutions, and were published in peer-reviewed journals. He had read them and they looked fine to him. That's all there was to it. Becker saw his role as giving medical judgment on the *significance* of the studies for human beings.

Harvey was outraged at this development. He said it was inconceivable that Becker could sit up there on the stand and not be willing to discuss the details of the studies on which his testimony was based. Harvey threatened to appeal, to go to court, to do whatever was necessary to insure that his client's constitutional right to cross-examine Becker and to test the quality of the evidence being offered against his client was not infringed.

This argument continued well into the third day, and by the time it was resolved—in Becker's favor—I think that whatever small favorable feelings he had for lawyers as a group were destroyed.

Simpson had told Matias, early in the scheduling of the hearings, that Becker would be available for four days' cross-examination. That was done at a time when we presumed the cross-examination would have some bona fide purpose. The power companies quickly accepted the tender of Becker's time and now we were stuck with it despite the fact that it was resulting in nothing but the alienation of Becker himself.

The third and fourth days were carbon copies of the first and second. Wallace suggested that Becker's work was nothing original, and that it seemed he was making a big deal out of nothing. Harvey kept asking Matias to throw out Becker's testimony because it was "the rankest form of hearsay" and "me too" testimony. But the Commission would certainly not let Matias do that, and Harvey knew it.

Becker left the stand at the end of the fourth day still saying the same things and feeling the same way about each point in his testimony as he did on the day he started. Beyond abusing him, the opposition had accomplished nothing. He was the only physician in the hearing; the only scientist who had founded and headed a research lab, performed animal research directly related to powerline NIEMR, and performed human experimentation. He had authored or co-authored more publications dealing with biological effects of NIEMR than perhaps anyone else in the world, and had helped found the field of science that deals with the interaction of NIEMR and biological systems. But the tools of the lawyers were used to wall him off, not to facilitate the search for truth.

PART III

CHAPTER 10

Playing Dirty

When the hearing ended I felt a tremendous sense of exhilaration. I knew I had gone through a baptism of fire in both science and the law, and I realized my life would never be the same as a result. But mostly I felt relief. I was out of jail, off that damn witness stand, and now I could look forward to the return of some routine and normalcy. At times during that eight days I'd wondered whether I could go back the next morning to sit there and endure it; several nights I had a terrifying nightmare, always the same, filled with threat and violence. Now, I thought, all that was over. But very soon after the tension of the hearing had dissipated, I began to realize that the real nightmares were about to begin. A reaction phase set in, and Becker's professional life, and hence mine, began to worsen.

One of the first indications that things might take a bad turn had actually come before I had gone on the stand. It came from the lips of Morton Miller. Right after Schwan left the stand, Harvey put Miller back up there to get him to change some of his previous answers, and in the course of his testimony Miller leveled a blast at me. He said that he had suffered an electrical shock when he examined my exposure apparatus during his visit to the lab (impossible—the power was off), and that with the help of RG&E engineers he had repeated my experiments and found that my results were due to artifacts (shocks to the rats). I had half expected such a fabrication, but what he said next was a surprise. He said he had been approached by Robert Flugum, an official of the Energy Research and Development Administration (ERDA, the predecessor of the U.S. Department of Energy), and offered money to do powerline-related research. Miller said that he told Flugum he thought there were no effects due to powerline NIEMR, so that such a search would be a waste of money. Besides, he said, he didn't want to spend half his life "looking for the needle in the haystack." But when someone asked Miller what he was going to do—whether he would take Flugum's money—Miller asked to go off the record to make a "humorous comment." "I follow the golden rule," said Miller. "He who has the gold makes the rules."

The only conclusion one could draw from this answer, it seemed to me, was that Flugum had plenty of gold to spend on studies that would be aimed at finding "no effects." So after the dust settled from the hearing, I looked into the matter, to find out who Flugum was and what ERDA was.

Some years earlier, Congress had split up the old Atomic Energy Commission into two separate agencies, the Nuclear Regulatory Commission—to regulate the nuclear power industry—and ERDA, which was mainly intended to promote and advocate energy development, particularly nuclear energy. Its very mission was to be pro-utility, and it was, unabashedly so. It thus had an interest in possible health risks from powerlines; it also had a *lot* of money.

Flugum was an electrical engineer who had worked for government and industry for about twenty-five years. I found an article he had written in late 1975—well after our testimony in New York had become public—in which he scoffed at the possibility of health problems from powerline NIEMR, and also suggested that, as with alcohol, a little exposure might even be beneficial. Flugum was also an advisor to Cyril Comar of the Electric Power Research Institute (EPRI), the man who had quashed the Gann experiments at Johns Hopkins. These two positions meant that he controlled most of the money available for powerline bioeffects studies.

I found out that Flugum had approached others in addition to Miller, making it known, in effect, that he had money for certain kinds of research—"inviting an unsolicited proposal" as it's called. Miller got the money he had been offered and more—several hundred thousand dollars—and began his experiments, presumably by Flugum's "rules." But even worse from my point of view, Flugum worked out deals with other investigators who were connected with apparently reputable organizations. Chief among the projects Flugum caused to be funded was one headed by Richard Phillips at the Battelle Pacific Northwest Laboratories in Richland, Washington. Battelle, a research-for-hire corporation, had hundreds of contracts with the power industry and with ERDA in energy development. Phillips, who for the most part was previously unconnected with the NIEMR issue, was awarded several million dollars essentially to repeat my experiments, but, I suspected, to find "no effects."

Phillips thus had a thousand times more money to spend on NIEMR experiments than I did. If Flugum's rules were to apply to Phillips's work, as the rumors and evidence available to me in 1976 suggested, then I was going to be met with overwhelming force in the battle over the existence and nature of NIEMR-induced biological effects. Phillips simply had too much money—more than *all* the investigators in the NIEMR bioeffects area had spent in the previous ten years. I tried not to be paranoid, but with millions to spend from ERDA, and millions available from the power companies, I was deeply afraid that the truth was going to be the truth according to Flugum and Phillips, and not the truth according to the facts.

*

As Miller's and Phillips's funding stars rose, Becker's began to set. In 1976 he had routinely applied for continuation of his research support by the VA. He was a Medical Investigator, a privileged status within the VA given in recognition of his past accomplishments, which included winning the VA's highest award for research. His job permitted him to concentrate on research and delegate most clinical duties to other orthopedic surgeons. Yet in June 1976 he was told by Dr. Marguerite Hays, his immediate supervisor in the VA office in Washington, that his medical investigatorship would not be renewed. Hays, who had begun working in the Washington office only around 1975, was bitingly hostile to Becker in her phone calls and letters in the weeks preceding her June decision, and he seemed utterly at a loss trying to explain or understand her attitude. She told him that the medical investigatorship program had been terminated by the VA—which turned out to be a lie—and that the scientific reviews of his proposal had indicated that, even if the program continued, he wouldn't have been funded because his work was of poor quality. We obtained copies of the reviews, and found that three were highly laudatory and that only the fourth, the briefest, was negative.

The tone, length, and structure of that fourth review—the comments were very general, as if written by a nonexpert, and it was only six sentences long—made me think that it was Hays herself who wrote the minority opinion.

However, Becker had been around for a long time and knew the ropes, so we survived Hays, at least for a while. Hays was in charge of only one of the two pots of money for research in the VA system. Becker massaged the grant proposal into a different format, submitted it to the individual who administered the second pot of money, and got it funded. He had lost his medical investigatorship, which meant that he had to resume his clinical responsibilities and seriously limit his time for research, but the laboratory kept going, and I kept going.

A little later in the year, however, we lost another government grant that, along with our VA support, was the main pillar on which the laboratory had been built: a grant from the National Institutes of Health. We had always had that grant, but within months after Becker and I became "controversial" we lost it, despite the fact that it had been the most productive of its kind in the USA, and had been well received by the NIH scientific Study Group that had evaluated it.

Another suspicious thing that happened was sudden trouble with some journal editors regarding our manuscripts. There is a certain element of luck and capriciousness whenever an investigator submits a manuscript to a journal for publication. Such problems are generally known and accepted by investigators, and we had had what I suppose was our share of them. But beginning around mid-1976 several unprecedented conflicts with journals developed. The one that incensed us the most, by far, was the conflict with *Science*.

In 1974 I had become intrigued by a form of high-voltage photography called Kirlian photography, which had been invented by Soviet scientists in the 1950s. The technique involved the production of fantastically intricate and complicated visual patterns of various body parts—fingertips, elbows, lips, for example—via the use of high voltages, but no other light source. Some American and Soviet scientists were claiming that the color photographs, properly interpreted, would yield diagnostic information about the physiological state of the organism; the color pattern or "aura" was said to be a biological "energy field" of some kind that reflected the health of the individual. There were many stories about Kirlian photography in the quasi-medical literature, and some claims that the photographs could be used to diagnose alcoholism, sexual arousal, and the presence of disease. The most spectacular claim was that of the so-called "phantom leaf effect." It was said that if a section of leaf was cut away and the remainder photographed by the Kirlian technique, the removed part of the leaf would appear in the photo.

The claims being made for Kirlian photography seemed far-fetched. But there was no law of science I knew that would preclude the effects, and I was fascinated by the possibility that the technique might have some meaning. I persuaded a medical-equipment manufacturer to provide equipment and expertise in helping me design and build the high-voltage circuitry required. Working part time, I spent four months designing and building the circuitry and seven months performing the experiment. In the end I concluded that the claims were wrong and that the "aura" arose from corona, a process that is reasonably well understood. I found that the "auras" from living and dead organisms were essentially the same, so long as the water content was identical.

At the end of March 1976 I sent a manuscript describing our experiments to *Science*. Several months later it came back, rejected on the basis of a single anonymous reviewer's comment, which was enclosed. The comment said, "Kirlian photography is one of the many crackpot methods generated by a lack of understanding of natural phenomena and a bent toward mysticism. Publication of this paper, which exposes and essentially demolishes the myths connected with Kirlian photography, may unintentionally, because of *Science*'s prestige, give a certain aura of scientific importance to the technique. I suggest that ignoring Kirlian photography is the attitude to be followed."

Kirlian photography, as it had been touted then, was a crackpot technique. I knew that because we did the experiment. *But how in hell did the reviewer know it?* The reviewer's attitude really angered me; I thought it was arrogant and ignorant. Kirlian photography had been getting a lot of press nationally, and now that we had done a definitive experiment that explained the technique and "demolished the myth," I thought it should be published.

Despite my irritation, I regarded the problem as routine. I had simply got a bad reviewer and needed to send my paper elsewhere. But one day in July, several weeks after *Science* had rejected the paper, I got a phone call from a physicist at Drexel University in Philadelphia. He told me he was a member of a group working on Kirlian photography and that he had heard about our work. We talked for about half an hour, and it became apparent that we had each done essentially the same experiment and reached essentially the same conclusion. He told me that he was sending his work to *Science* for publication. I told him about the problem we'd had with our article, and he said that his group wouldn't have the same problem because "we know somebody at *Science*." He told me that the Kirlian work at Drexel was being funded by the Advanced Research Projects Agency in the Department of Defense to the tune of about $300,000. That was about 500 times more than our experiments had cost.

Sure enough, the October 15 issue of *Science* contained the article on Kirlian photography by the Drexel group. The article reached exactly the same conclusions that we had reached and, I thought, wasn't nearly as well done. The Drexel group had robbed us of our precedence and had gained access to the enormously large audience that follows *Science*.

Becker was furious. This was one of the most unethical things he had seen in all his years of research. He sent a letter to *Science* complaining about what had happened to us and received a reply from the editor, Philip Abelson, telling Becker that he had been "overly quick in coming to a slanderous conclusion," and demanding an apology. This answer further incensed Becker and he wrote to the President of the American Association for the Advancement of Science, *Science*'s parent organization, and to the publisher of *Science* seeking a full inquiry into the affair, and an appropriate remedy if it was found that we were correct. The publisher, William D. Carey, said that he had investigated the matter personally and that he believed the Drexel group—who were denying that my phone conversation with them ever took place—and that he didn't believe Becker.

But that wasn't even the point. The point was that our work had been rejected simply because it dealt with a topic that the editors considered "unscientific," and yet when someone else studied the same topic, it *was* sufficiently scientific. Carey did not respond to that point at all.

The *Science* episode hurt us. It robbed us of recognition and priority, both of which would have been useful when we applied for grant support in other areas. It also lowered our expectations of ethical or fair conduct in the practice of science. But most important, the episode was intimidating. Our laboratory had always been fast-paced and exciting, and if we made people angry along the way, we simply never stopped to dwell on it. Everybody has somebody who doesn't like him. But we had never felt before that we had bullseyes painted on our backs.

In April 1976, one month after I had sent in the Kirlian article, the charges Becker and I had made that the NAS Sanguine committee was rigged were reported in *Science*, together with the official denials by NAS. The article was written by a well-known writer and it was eminently fair to both sides, I thought. Paragraph by paragraph it described the background of the situation, the various positions, and the arguments on each side. But the *gestalt* of the article was that two obscure upstate New York scientists were criticizing—severely criticizing—the integrity of the National Academy of Sciences. If I had read such an article about two people I didn't know, I guess I would have wondered who they thought they were.

*

After the New York hearing was over, NAS contacted Becker and asked him to be a "consultant" to the Sanguine committee. If Becker had accepted this offer, he would have been in the bizarre situation of being an advisor to Schwan, Michaelson, and Miller. Unlike the thrust of the power-company approach to Becker, which I could only characterize as abusive, the NAS offer was designed to appeal to Becker's sense of importance. He was to come in late Spring for a few days to Wood's Hole, Massachusetts, where he would meet with the Committee, hobnob with relatively important people in the scientific community, and be paid $125 a day for his trouble. NAS had contacted virtually all NIEMR investigators in the country (except me) and asked them to be "consultants" to the Committee.

The invitation to Becker was a setup that even a blind man could see. All a consultant got was his name listed in the front of the final report—no one was required to *listen* to him or *believe* him. Even worse, if the Committee criticized his research in the report, then the listing of the investigator's name as a consultant would make it seem as if the investigator himself actually concurred in the criticism; at the very least, it would appear that the Committee had gone the last mile to learn about the details of the investigator's experiments. I advised my colleagues generally not to take part in this charade; however, many did and I think most of them regretted it. NAS tried almost everything to get Becker to be a "consultant." They sent one of their staff people to Syracuse to see him, they called him, and they wrote him. At first, Becker said he couldn't go because of the demands of his practice and his experiments, but then NAS told him that wouldn't be a problem because he could consult with the Committee via a telephone hookup from his office. The process continued in that manner for several months—an offer by NAS, a respectful declination from Becker, and a counteroffer to accommodate Becker's stated reason for not taking part. Finally NAS gave up the game.

CHAPTER 11

All Over Again?

During this time, the hearing began to heat up again. The power companies had appealed to the Public Service Commission, saying that they hadn't had enough time to cross examine me (!)—they wanted me to go back on the stand for an indefinite period, maybe weeks, or months. It was an outrageous demand. The Commission, in its typical role as compromiser, averaged the staff's position of zero and the companies' position of infinity and ordered me to go back on the stand for two more days. The Commission did not *ask* me to go back; they didn't appeal to my sense of duty or patriotism or any other grand ideal. They just ordered me. I sat there listening to Simpson telling me by phone about the Commission's "order," and all I could do was make obscene gestures.

Now, I was a free agent, completely free, and I could walk out of the front door any time I pleased. There was nothing the damn Commission could do about it. They couldn't legally order me any more than they could order the man in the moon. I figured I had done enough.

But after a few days things began to look a little different. For one thing, I saw the text of the Commission's actual "order." It said that I was a "very important witness" and it severely criticized Harvey for wasting my time, and for not examining the actual NIEMR studies on which my testimony was based. In the two additional days, it said, Harvey was not to make that mistake again. Well, that mollified me a little. Another thing was that Simpson convinced me that if I didn't go back for the two days, the power companies would have a strong case for throwing out my entire previous testimony. I really didn't doubt that Matias would do that—I was convinced that he hated my guts—and what might happen after that was very unclear. Would Rheingold allow Simpson to appeal such a decision to the Commission? What would the politically sensitive Commission do? If the matter ever got to the courts, what would they decide? I knew Simpson was giving me the best advice he could, given all these imponderables.

Either I submitted, or I risked seeing everything I'd done so far go right down the drain.

So, in September, I went back on the stand for two days. Simpson, not wanting to be accused by Harvey of interfering, hardly spoke. Harvey, who was in effect under the Commission's thumb, confined almost every question to the NIEMR experiments. As for me, since I was utterly convinced we had already won the case on the merits, I answered monotonously and without passion. The two days were psychologically stressful, and a complete waste of time. We were merely going through the motions.

*

In 1976 I was thirty-six years old and I had a Ph.D. in biophysics, twelve years' experience doing research in the biological aspects of NIEMR, a degree in law, and an embryonic law practice. I had a family and a house with some nice land, and I knew how to do a lot of practical things like carpentry and growing tomatoes. Thanks to my family upbringing and to the Jesuits at St. Joseph's, I thought my life had solid philosophical underpinnings. I saw life as pretty much a linear affair—beginnings, endings, new beginnings. I certainly didn't think of myself as naïve, especially after what I'd just been through. But toward the end of 1976, after I had left the stand for the second time, I got one of those jolts that strike at the heart of the way you see the world. The Commission ordered that the entire hearing was to be done over again.

New testimony, new cross-examination, maybe new witnesses. I couldn't believe it.

Why?

Since the PASNY line was already being built, it was actually the RG&E/NiMo line that was the real point of interest. But the hearing would also affect other lines planned for the future. What seemed to be happening was that PSC was stringing out the decision process until a political consensus on high-voltage powerlines emerged. For instance, would consumers pay more money for electricity in order to assure that people living along powerline rights-of-way were protected against health risks due to NIEMR exposure? That judgment, and the political fallout associated with the location of the PASNY line, were both quite unclear. PASNY was cutting down trees and bulldozing the land, a process that was producing a public outcry of increasing volume from the farmers and other landowners who were directly affected. How far these people would go—whether they would hire lawyers and fight, whether they could generate any political clout—was unknown.

Thus the power companies, which seemed always to want to rewrite the record, and PSC, which was being blown about on the sea of indecision by the winds of politics, apparently had a common interest in continuing the agony of the hearing. So the power companies asked and PSC ordered it all to be done over again.

The rule was to be that all testimony from the first hearing would be combined with that from the new hearing, and after that the Commission would make a final decision. I don't think anyone, especially Harvey, expected that I would be involved in the new hearing as well. He had never had an answer to his oft-asked question about my participation in the first hearing—"Why is he doing it? What's in it for him?" Surely he couldn't see the sense in my continuing.

For my part, I felt as if the Commission might *never* make a decision, that the hearing might go on forever, but that in any case my testimony was part of it. I had done my job. I couldn't conceive of anything that would make me voluntarily go back for more.

But that was before I learned what the power companies had in mind.

The PSC rules required each side to tell the other the names and qualifications of witnesses they intended to call. Accordingly Harvey notified Simpson that, in addition to the witnesses from the first hearing, he was going to call new ones. The one who interested me most was Henry Hess. He worked for a consulting firm that wrote environmental impact statements for power companies. I learned that Harvey had given this man copies of my raw data. Hess was supposed to have done some sort of computer analysis of it and—surprise!—concluded that my statistics were faulty and my experiments had actually *not* shown that NIEMR affected rats and mice.

Now, the power company witnesses—the best that money could buy—had self-destructed on the stand, and it was certain that their analysis and reasoning would have no decisional impact. Thus, it was probably foolish to be concerned about the words of a power-company employee testifying for his bosses, especially when he had virtually no research experience. The thing to do was pull out, and let the PSC people have their chance. Matias would be balanced by Simpson; it was time to let them settle the matter.

What happened, however, was that I became afraid, afraid that Hess might be believed unless I was there to point out every lie. It was one time when I think my legal training actually hurt me. I envisioned a kind of super bookkeeper up in the sky who put a check mark under a column in a ledger that said "Charge against Marino"; I had to be around to provide the check in the other column that said "Marino's reply." I felt my reputation was at stake, and that if I didn't stay it might look as if I was hiding from some new "expert" who had found my Achilles' heel. The hearing had become very public; we were constantly being interviewed by the press. For me to leave now could look very bad.

There was another thing as well. I really wanted to get Harvey and Wallace. I had more information now about how the power industry had behaved throughout this controversy, and I wanted to rub their noses in it. I know that wasn't a good basis for a decision, but as time went on, and the aggression against me and Becker increased—from all quarters, it seemed—logic took a back seat to anger and frustration.

Simpson thought I should stay in the new hearing too, because my new written testimony would contain nice capsule summaries of all the weaknesses of the power-company experts' testimony and thus provide a kind of road map through the hearing transcripts, which were already about twenty feet high. In addition to that, and to being on hand to protect myself when people started calling me names, I would also be able to describe even more NIEMR studies which tended to show that the powerlines would be a public health problem. Since my first testimony, thirteen additional studies had been published or otherwise made available. So now I had forty-five.

*

If I was going to stay, I thought it only fair that I should have as much information as the adversaries. If Miller, for example, had a NIEMR study that showed effects, and I didn't, then sure as hell he wasn't going to give it to me voluntarily or even mention it in the hearing. By and large, that was no real problem because he simply didn't look for the work of others. After the first hearing was over, for example, he wrote me for copies of sixteen of the thirty-two NIEMR studies that I had described in my testimony. The man had said that the powerlines would be safe not only without doing any research himself, but without even reading about the research others had done.

But there was the possibility that somebody had found and given him articles of which I knew nothing. The one strong indicator of this possibility involved Miller's role on the NAS Committee. NAS obviously had access to information in foreign journals, government reports, etc., that I lacked, and during the first hearing, when I was delivering my colloquy on the Noval study, Miller had blurted out that he had "three feet" of material from NAS. I told Simpson that *I* wanted copies.

Simpson asked Harvey for them and Harvey turned him down flat. Simpson then wrote to NAS directly and NAS turned him down flat. Simpson then told his big boss, Alfred Kahn, Chairman of the Public Service Commission (the man who, a short time later, became President Carter's kingpin in the war against inflation) and Kahn wrote directly to Handler on Simpson's and my behalf requesting access to the material.

In October NAS told Kahn that they would not provide PSC with a copy of the material that they had given Miller. Enclosed in the letter to Kahn was a photocopy list of NIEMR references which Kahn was told might be helpful. The list consisted mostly of a photocopy of the references from *my testimony.*

I thought that NAS's treatment of Kahn was contemptuous. So did Becker, and he sent them a letter explicitly terminating all contact with them over the Sanguine Committee matter because, he wrote, it was completely inappropriate for an organization whose primary goal was scientific truth to refuse scientific information to government officials who needed it to make good decisions in the public interest.

*

In November 1976 Simpson mailed my new testimony to the power companies. Becker, who was by now disgusted with the whole process, didn't write any more testimony because he felt there was nothing more to say.

Schwan's new testimony was oddly weak and defensive. I think his reaction to the entire adjudicatory process was much like Becker's: he must have felt that this was *not* the way to make a decision. Miller's and Michaelson's testimony was an "analysis" and a condemnation of *all* NIEMR studies that had found effects as incompetent or otherwise inadequate; it seemed to me not the slightest bit believable.

Hess's testimony was something else. What he did was to take my raw data, massage them to suit his purposes, and then conclude that my statistics were faulty and what my experiments *really* showed was that NIEMR was safe. Hess generated hundreds of pages of computer printouts that seemed to me to have very little to do with my data. Hess said that he had analyzed my data, but by completely misanalyzing them, what he had actually done was to make up his own.

So I set out to write a report to refute their report (by Hess) that had been intended to refute my report (my November 1976 testimony) which had been intended to refute their reports (by Schwan, Miller, and Michaelson) which had been intended to refute my report (my December 1975 testimony) which had been intended to refute their 1974 report (by Kanu Shah in the original R&E hearing). Life, for me, was no longer linear; it was circular, or at best, pogo-stickish.

While I worked on my reply to Hess, Harvey asked Matias to throw out almost all of my second testimony...and Matias did. But Simpson appealed to the Commission, and they overruled Matias and put it all back in.

As I finished working on the reply to Hess, Matias got word of it and told Simpson that he wanted to see it before it was sent out to the power companies. After Simpson showed it to Matias, Matias said that Simpson *couldn't* mail it out. Matias felt that what Hess had said made sense to him, and he didn't see any need for me to answer Hess. Again Simpson appealed to the Commission, and again the Commission overruled Matias and ordered Simpson to send out my report and Matias to admit it into evidence in the hearing.

With Matias being repeatedly overruled by the Commission—a most unusual rebuke for a judge—I suspected that his anger at me was reaching the boiling point.

CHAPTER 12

Citizens' Revolt

The landowners and citizens' groups of northern New York had been stunned that PSC would let PASNY build its line before the hearing had concluded. As far as they could see, the decision had been made strictly on economic grounds and because, as the one dissenting Commissioner charged, his colleagues had been "terrorized" by PASNY. The landowners filed suit against PSC and PASNY to have construction stopped.

But it was not in court that the most significant actions took place. Here is an account of what happened next, according to a letter written by the farmers of the Fort Covington/Bombay area of northern New York:

In 1976 they said, sure enough, the line was going to come right over our farms and an easement was to be acquired by PASNY. Their land agent set the tone for PASNY's dealing by telling us we might as well sign, that the line was coming through whether we wanted it or not. He told us not to get a lawyer because they had the best lawyers and never lost. We wrote letters to the Governor...he did not answer. We spoke to our Congressman...he said never to write him again. We didn't sign, though, and we did get a lawyer and prepared a suit to protect our land. At about this time, October, their contractors came onto our land and started to put a gravel-covered culvert in a low ditch. It was wet...so they said they'd use a tracked vehicle to haul the gravel through the meadow. That was too slow...so they drew ten-wheel gravel trucks through the meadow with a bulldozer and put in an eighteen-inch pipe where their own prints called for a forty-two inch culvert. The water ran over the gravel embankment like a dam. At this point it was clear we were dealing with wreckers rather than constructors, and twenty of us gathered at the meadow to stop their equipment.

Thus began the citizens' guerilla war against PASNY. In response to this threat, PASNY went to court and got a blanket injunction against anyone who tried to interfere. The landowners appealed, but the judge wouldn't hear their case. Frustration began to mount. Not long after a large rally against the line in Fort Covington, PASNY crews razed June and Glenn Black's apple orchard—after promising they were only going to "top a few trees." June Black, Doris Moeller, and Georgette Lauzon refused to allow the crews back on their land, were arrested, and spent the week before Christmas in jail.

This act of defiance galvanized the resistance. In early January 1977 a determined group of landowners and their supporters, including several Native Americans from the nearby St. Regis Mohawk Reservation, prevented PASNY crews from entering the farm of Harold and Stella Barse. After a week's tense confrontation, marked by some near-violent skirmishes, state police arrested twelve people and the crews finally cut down the one-hundred-year-old live elm tree they were after. The stand-off at the Barse farm received widespread news coverage and generated new support for the protest.

About the same time Governor Carey said in a news conference that the 1977 winter fuel emergency in New York should resolve anyone's doubts about the need for the 765,000-volt line. Carey didn't point out that the line would carry imported Canadian power only during the *summer* months. Moreover, his statement confirmed the growing sense that he had no concern for the North Country citizens and their farmland, and that he was being fed erroneous information by James Fitzpatrick, Chairman of PASNY. The protest groups thus began to organize more seriously now for what looked to be a long struggle against the state.

In late January the *New York Times* ran a feature article on the protest. About that time a statewide coalition of citizens' groups formed to support the North Country resistance, and a major political strategy conference was held in Holland Patent, near the southern end of the line. In the field, interference with construction work continued and there were further arrests.

PASNY responded aggressively. Its public relations director—his name was Spieler—charged that the protest was led by outsiders. Farmers were offered paltry sums for easements and given no appeal. Earlier PASNY had threatened St. Lawrence County with a suit if it passed a proposed agricultural district law to protect local farmland; now, other state agencies were pressured into denying requests for public hearings on the agricultural impacts of the line. When local officials and citizens traveled the six hours to Albany, to testify before a legislative commission about PASNY's practices, Fitzpatrick himself attempted to filibuster so they would have no chance to speak. At each public meeting or protest, PASNY agents could be seen taking photographs and noting car license numbers of the protestors. But the citizens kept up the pressure, and PASNY's public relations problems steadily worsened.

The farmers in northern New York have a reputation as a conservative, independent-minded, and self-sufficient people. Their relation to the land goes back to the time of the Revolution; they are the last people on earth one would expect would throw themselves on the ground in front of 18-wheelers, or chain themselves to trees that some government agency wanted to cut down. The only way to outrage that kind of a person that much, I thought, was to treat them with utter contempt, and that's exactly what PASNY and its lawyers and agents did. PASNY, it seemed to me, was a rogue agency that had been turned loose on really defenseless people...and by a liberal Democratic governor. (Liberal Democrats, I had always believed before, cared about "people" as opposed to Republicans who only cared about "business.")

The people in northern New York didn't mean a thing to me personally, as individuals. I didn't know any of them. I had no relatives who lived there, and I had no interest in any farm, or business, or anything like that. People I didn't know from Adam would call me and tell me the most outrageous stories, and it just made the hair on the back of my neck stand up. One lady told me about her 86-year-old mother who owned several hundred acres through which PASNY's line was slated to go. The lady was partially blind and crippled, and lived in a nursing home. PASNY's land agent visited the nursing home three times in an attempt to get her to sign a paper that would facilitate their construction of the line across her land. Finally, unable to resist any longer, she signed the paper, but at the top she wrote, "I didn't want to sign...they made me do it."

There was only one other time in my life that I had become as angry as I did with PASNY. It happened in 1970 shortly after my wife and I, and our then infant son, moved onto the five-acre farm just west of Syracuse. There was a small stone quarry nearby, and after we moved in the quarry owners petitioned the town for permission to expand operations. Even before they got the permission, they increased their activities: there was a lot of dust and noise, there were explosions to free the rock, and many, many more trucks to carry away the crushed stone. I went to the public hearing to represent my family and neighbors. I wasn't a lawyer then, but I wasn't exactly tongue-tied and I expected to have my say.

But things didn't work out quite the way I expected. Just before the hearing began, the stone company entourage came into the room. There must have been ten of them: lawyers, "expert" witnesses, company officials, even a stenographer. The company presented its case first. Their witnesses said that they had measured the sound levels on property adjacent to the quarry—such as mine—and found that they were very low, comparable to the background level in a quiet office. In fact, they said, the chirping of the birds was much louder than the sound created by the quarry. Explosions from the quarry could knock you out of a chair, but a company witness testified that the vibrations from these explosions were virtually undetectable beyond the company's property. One "expert" on sound was brought in from North Carolina State University to testify that this was an ultramodern stone quarry and, by implication, only quacks would oppose it.

The hearing droned on and on, and my blood began to boil. But I didn't know what the rules were, I didn't know when I would have a chance to speak, whether I *would* have a chance to speak, whether I could question the witnesses. The whole thing was tremendously frustrating; the "experts" just kept talking bullshit. Finally, something snapped, and I just stood up and started arguing. I don't remember exactly what I said but I do remember calling the chairman of the zoning committee a tyrant as he banged his gavel harder and harder so it could be heard over the cheers and applause of the audience. I also remember him crooking his finger at the two policemen in the back of the hearing hall and then pointing to me.

My performance, however cathartic, was a disaster in practical terms; I never got a chance to present any evidence or give my opinions or do anything that would give me half a chance of convincing the Board that I was right. I self-destructed, got tossed out of the hearing, and the company got its zoning change.

But this time I was a lawyer, and in my best legal prose I wrote a letter to Governor Carey in the hope that he really didn't know what was going on in northern New York, and that once he did, he would do the right thing, or at least look into the matter. I wrote:

The most basic issue in the PSC Hearing is whether exposure to the field of high-voltage lines constitutes a human health hazard. Notwithstanding that the issue is presently sub judice, recent events compel us to inform you that the 765,000 volt transmission lines as presently designed endanger public health, and to request your urgent assistance in halting construction of the Power Authority's 765,000 volt transmission line until its health hazards are properly considered....

The basis of the Commission orders permitting construction by PASNY are faulty. The Commission has no tenable basis in the record at this time to permit PASNY to begin construction. It seems clear to us, as intimate participants in these proceedings for almost three years, that the action of the New York legislature in the spring of 1976, in preparing to pass a bill declaring PASNY's transmission line to be safe, was a significant factor in the Commission's decision to administratively approve the line....

We respectfully request that you take immediate steps to halt construction of PASNY's line until the health issues have been resolved. The present policy of piecemeal certification will obviously result in a vested economic interest on the part of PASNY which will preclude any decision, executive, administrative, or judicial, to deny PASNY the right to energize the transmission line, notwithstanding its hazards to human health.

Though my language was more professional this time, the results were no better. Carey replied that PSC was handling the matter just fine, and that if I didn't like that, I could sue PSC.

*

In February 1977 our feud with NAS over the Sanguine Committee became more public. Months earlier Dan Rather of the CBS show "60 Minutes" had come to the lab to interview Becker about the Sanguine matter. After the interview I had spoken to Rather and to the show's executive producer, Richard Clark, about the powerline controversy and what was happening in northern New York. Clark had thought it a good story for "60 Minutes," and said that he and Mike Wallace would return for an interview. In February, a few days after this interview was taped in our lab, Becker's Sanguine interview with Rather went on TV. (© by CBS, Inc.)

Dan Rather: You may have heard of it—a Navy project called Seafarer. The original name was Sanguine. It's a $700,000,000 submarine communications system that for the past decade has been a very expensive idea in search of a home. In every part of the country where the Navy has set foot talking Seafarer, there has been an uproar. Homefolks, politicians, scientists—they all turn out to have at it. Will Seafarer ruin the scenery? Will it be a major target for enemy warheads? And most important, what about reports that it could be harmful to the people who would have to live with it?... Captain Charles Pollack is the man in charge.

Capt. Pollack: The antenna would have about 2400 miles of antenna cable. If you draw a line around the extremities of that antenna-arrayed layout, it would encompass about 4000 square miles.

Dan Rather: So, somewhere in the good old U.S. of A., Pollack has to string out 2400 miles of antenna cable, buried a few feet underground. It would look something like this—a pattern resembling loose strings in a tennis racket. The intersecting lines

70

would be about 3½ miles apart, and the whole thing would cover 4000 square miles of field and forest—some of it along existing right-of-ways, like roads and powerlines, some of it through newly cleared paths. Is it safe?

Capt. Pollack: Yes, absolutely.

Dan Rather: Absolutely? Well not to people like this scientist. Are you telling me there's a possibility that electric current, generated in a fashion such as this, could possibly cause heart disease and/or stroke?

Dr. Becker: Yes.

Dan Rather: You have to know that that's a mind-blowing thought for a lot of people, including me?

Dr. Becker: I'm aware of that.

Dan Rather: Dr. Robert Becker is Chief of Orthopedic Surgery and a medical investigator for the Veterans Administration in Syracuse, New York. We have to pause here for a bit of explanation. Historically, the scientific community, almost in its entirety, has maintained that, to be harmed by electricity, you had to be shocked or burned; that the low-level doses surrounding us most of the time—from electrical appliances in the home, from power transmission lines or from the Navy's Seafarer project—could do us no harm. That's the Navy's argument. Now, are you telling me it's fair to say, accurate to say, that a housewife is exposed to more low frequencies in her home in the course of doing her day-to-day chores than she would be from seafarer?

Capt. Pollack: Many, many times more.

Dan Rather: You're certain that is a scientific fact?

Capt. Pollack: That is a scientific fact.

Dan Rather: Dr. Becker wouldn't disagree with that. What he'd say is that you may not be safe, even in your kitchen. For twenty years, he and his staff have been experimenting on the effects, if any, of low-level radiation on living things. He is one of a small, but growing group of scientists around the world who are turning up information making them believe that low-level electrical fields do affect us. For instance, using very low voltage currents, he has made broken bones that wouldn't heal by themselves grow together again. And like most scientific discoveries, it's a double-edged sword. If those carefully controlled low-level currents can heal bones, well, it makes people like Becker wonder about uncontrolled electrical fields from household appliances, powerlines and Seafarer.

Dr. Becker: I was a member of the first ad hoc committee to evaluate the biological studies that were performed for Project Sanguine. And I most certainly sat there and listened to several studies that had very definite effects, yes. Animals that are exposed grow at a slower rate than control animals. A number of projects have shown this to be true. The second area in which definite effects do appear is that exposure to this type of field seems to produce stress.

Dan Rather: Is it true that the Navy repressed that report for better than two years?

Dr. Becker: The Navy did not disseminate the report widely.

Dan Rather: This is the report that he's talking about. It was done at the request of the Navy by a group of top scientists. They reviewed experiments performed on possible effects of ELF radiation on living things and raised some red flags. That was in 1973. The report finally got out a year ago.

Capt. Pollack: Now, in looking at some of the early experiments, there were effects noted. There were differences of opinion among—among the scientific people, as to whether those effects were significant. There were also differences of opinion as to

71

whether they were deleterious. It's our position now, and I fully support that position, that we have not seen any significant deleterious effects that can be attributed to a Seafarer system....

Dan Rather: Of most concern to people, though, were those reports of what the grid might do to them. The Navy countered: More scientists were following up the original studies. But the folks here had heard stories about what happened to some of the original scientists.

Dr. Becker: We know of, I believe it's five, specific projects in which positive results were obtained, when the projects were terminated and the money just disappeared. There was no more to continue the work.

Dan Rather: Now, is this a definite pattern?

Dr. Becker: It appears to be.

Dan Rather: That when a study begins to find that there may be adverse effects, that the money dries up?

Dr. Becker: Not even adverse effects. *Any* effect.

Capt. Pollack: What we had to do is try and determine which of the research efforts appeared to be the most fruitful, where we should apply the money and where we should apply the resources in order to get the best overall understanding.

Dan Rather: Meanwhile, the Navy has called the National Academy of Sciences to oversee and evaluate further experiments. And the NAS Committee has issued a one-sentence interim report saying that, so far, they think Seafarer is safe. But Dr. Becker isn't impressed. Some members of that NAS panel have previously testified publicly that radiation, similar to that of Seafarer, isn't harmful. And Becker maintains it would be awkward for them to change their minds in public.

Dr. Becker: For example, if a person has already publicly gone on record that the voltage field from a transmission line, a million times stronger than that from the Sanguine antenna, is harmless, then obviously he cannot do an about-face and say the Sanguine antenna may be harmful. So that a number of people on this committee, I would feel, have a pre-bias.

Dan Rather: Is what you're trying to say that we're playing with a stacked deck?

Dr. Becker: I think so, yes.

Immediately after the program was aired, NAS president Philip Handler sent a letter of protest to the president of CBS, demanding a retraction of the charge that the committee was stacked. Such charges, he said, were "laughable" and "intolerable." Published in the Detroit *Free Press*, Handler's letter suggested that Sanguine was safe, even though the committee, which was supposed to be evaluating that question, had not yet issued its report.

*

In preparation for Mike Wallace's interview with me, Richard Clark had asked me to send him whatever literature I had available about the background of the issue, including the possible risks of powerlines. I wrote a letter explaining things as I saw them, and I assumed that the power-company experts had done the same. But when he came to my laboratory I learned that they had not sent him anything, and had refused to be interviewed. Wallace called Robert Harvey from my office and asked him to make his witnesses available so that "forty million Americans could learn what the power companies had to say." Harvey first said that he had to confer with somebody, then he called back and told Wallace that the witnesses were free to do anything they wanted—that they weren't controlled by the power companies. Wallace called Michaelson, but Michaelson refused to be interviewed because he said he was "too busy." Miller also refused because, he said, he never discussed his scientific views in the press—only in scientific publications. Wallace finally located someone at the University of Rochester—a minor witness in the hearing, named Edwin Carstensen— who would defend the power-company position on the show, and after he finished the interview with Becker and me he immediately drove over to Rochester.

During the interview in Rochester Wallace asked several questions while brandishing the letter that I had written for him and Clark. When the interview was over Harvey called Simpson and demanded a copy of the "report." I told Simpson I had no copy, and the power companies were infuriated. They claimed that I had poisoned Mike Wallace's mind with my propaganda, and if they could get a copy of it they could show that I was prejudiced, biased, and not believable.

*

Cross-examination of my November 1976 testimony and my reply to Hess took place in March; the solons at PSC had decided that it should be three days—no more, no less. This time there were many people there from northern New York, and that helped a lot. Matias seemed miserable, almost depressed. The people from the North Country no doubt remembered his flip attitude from the original PASNY hearing almost four years before. Also, he was back in Syracuse again, and I guess, like me, he figured that this damn case might never end. He had already been overruled twice by the Commission about my testimony, and I know he was angry about my letter to Carey which, after several months without any response, I had made public. He told me that, as a lawyer, I should have known not to write letters about things that were still involved in legal proceedings.

For the three days I was on the stand, the thing that the power companies seemed most interested in was not my actual testimony, or my research, but my letters to Carey and Clark.

Frank Wallace, PASNY's lawyer, was the one most concerned with the Carey letter, and we talked about it for what seemed like half the time I was on the stand. Wallace was shocked ... he was outraged ... he was scandalized ... he couldn't *believe* I would have the audacity to do such a thing. He went on and on implying that I had misled Governor Carey, that I had exaggerated, that I had lied. What really ticked him off was that I had told Carey that Wallace hadn't bothered to show up for the hearing on two separate days. That's an extremely dangerous thing for a lawyer to do, particularly if he bills his client for those days. Wallace tried to get me to admit that I had made a mistake; but I had absolute proof that he wasn't there on the days that I said he wasn't. But I was like a fly on flypaper; I was there for three days and Wallace had to do something to rehabilitate his image. I understood that, and I just waited for the Big Wind to blow over.

It was Harvey who wanted the "60 Minutes" letter, and Matias agreed with him that it was crucial to the power companies' position, and that they ought to have it. There were endless discussions about it, with the dialogue

73

alternating between accusations that I must have something to hide because I wouldn't provide a copy, and evaluations of various strategies to get a copy directly from the "60 Minutes" people.

After almost four years of the most complicated legal maneuvering that had ever taken place in a PSC hearing in New York, after tens of thousands of pages of testimony, endless days of cross-examination, and hundreds of documents, it had come to this: that for the power companies, a three-page letter I had dictated off the top of my head in thirty minutes was the key to the whole case.

A frenzy set in. First the lawyers wanted to go to my office and search it for a copy of the letter. Then they decided to call Richard Clark in New York and ask him for a copy. One lawyer said that wouldn't do any good because I only had this one day left on the stand, and they wouldn't be able to get it to Syracuse from New York City in time to use it against me. But Harvey said he could solve that problem because he had access to machinery which could send photographs over telephone lines. So, if they could get a copy of the report in New York, they could telephone it to Syracuse. Wallace said that Matias shouldn't *ask* for the report, he should *subpoena* it. Matias said he didn't think *that* was a good idea because he didn't know if he had the power to subpoena anything. In the end, they decided that they would *ask* Clark for the report.

They called his office, and were told by his secretary that he was on vacation in Alaska. They asked *her* for a copy of the report, but she told them she couldn't do that; she said she'd be glad to tell him about the situation when he returned in two weeks. The lawyers said that their mission was very important, like a matter of life and death, and they had to talk to Clark as soon as possible. The secretary said that she would do her best and, for the most part, we twiddled our thumbs for the next several hours. Then, toward the end of the day—and freedom for me—a messenger came into the courtroom and said there was a long-distance phone call from Alaska. Simpson was chosen to represent Matias and request the report on behalf of the Public Service Commission. Simpson got on the phone, identified himself, and told Clark what Matias wanted. Clark asked Simpson what kind of an idiot Matias was—that he should know the report was part of material gathered for a story and that it was privileged information. In his best lawyer language, Simpson tried to do his duty—to impress on Clark the utter necessity, in Matias's view, for a copy of the report to be made available in the interests of the people of the state of New York. Clark listened, and listened, and finally told Simpson: "Tell him to go f--- himself."

Simpson came back into the courtroom and, on the record, Matias asked Simpson for Clark's reply. Simpson said, "Mr. Clark respectfully declines to make the report available on the basis of journalistic privilege."

The power companies were no better off at the end of the second hearing than they had been at the end of the first. The way I saw it, they simply took a chance that I wouldn't be around for the second hearing and thus that their experts would have free rein. When that didn't work out, all they could come up with was the pathetic attempt to get the "60 Minutes" letter that they hoped would be my undoing. It was pathetic because I had written the report anticipating that it might ultimately be made available to the power companies: it contained no words that I needed to take back, and

I would have been quite comfortable to read the whole thing on national television.

When I left the witness stand in March—for the *last* time—the "legal" phase of the hearing ended and everything was now up to the Commissioners. The only unknown that remained was the political clout from the publicity that could be generated by the protest in northern New York.

*

By the end of March, the protest had gathered great momentum, and over 1000 people rallied in the tiny village of Edwards, calling for a halt to construction of the PASNY line. The march was covered by the *New York Times*, the Associated Press, and the *Daily News*, and shortly afterwards "60 Minutes" aired its piece on the protest, including Wallace's interviews with Becker and me.

Despite the growing support for the protest, Governor Carey was adamant about the line. At a press conference he said: "That line has been cleared with all the environmental authorities and the PSC after a very careful review of the process ... those who are being arrested, as I understand, are not really the indigenous persons who own that land. They are, frankly, trouble-makers, who, after the review process and satisfying the environmental requirements, are still doing their thing." When a reporter pointed out that the hearing was not yet concluded, Carey asserted again that the line had been "cleared." The North Country groups angrily responded in a press conference of their own that the Governor's remarks about outsiders and troublemakers were either the result of misinformation from PASNY, or outright, knowing lies. Three members of local groups traveled to Albany and confronted Carey face to face, and five days later he retracted his remarks. Then, two weeks later, it was announced that Fitzpatrick would retire from the Chairmanship of PASNY. To replace him Carey named his personal friend Frederick Clark, an Albany banker.

But during a subsequent political visit to the North Country, Carey once again leveled the charge of outside agitation. One of the protesters, he said, was from Pennsylvania. Next came an attack from Carey's Commissioner of Commerce, John Dyson, who called the protestors "screwball farmers." (Dyson would later replace Clark as Chairman of PASNY.) That comment brought down on Carey not only other blasts from the North Country but from farmers and farm organizations throughout the state.

The legal maneuvering in court continued for a while but it was really a case of a gnat against an elephant. The farmers had scant financial resources, whereas PASNY's were virtually unlimited. It had all sorts of advantages actually written into the law which made it difficult for *anybody* to fight them. The landowners did get an injunction against PASNY, but New York law was that the injunction was automatically stayed as soon as PASNY filed an appeal. Thus, PASNY was free to build the line during the appeal, which would not be decided until *after* the line was operating.

In April I visited the North Country for the first time to deliver talks at two colleges. The second of the two talks was open to the public and the auditorium was packed. The emotional temperature in the hall was very high; the people cheered and clapped and gave me standing ovations. The only thing that took the edge off the evening were the PASNY agents who sat in the front row and tape-recorded everything.

It wasn't long after my visit that some of the very same people who were in the hall were sentenced to jail for their earlier activities against PASNY. As the defendants were brought before the judge to be sentenced, they were each given an opportunity to make a statement. If the language of these people—some educated, some not, most simply trying to protect land that had been in their families for a long time—didn't move those who heard it, or who read about it in the newspaper, then I didn't think anything could:

Stella Barse: I plead guilty for trying to stop experimentation on human life.

John Lauzon: This damn line is wrong altogether. If PASNY is going to wreck all the farmland, we might as well all be in jail.

Alan Casline: What I've seen has really changed me. I committed no illegal act but somehow I end up being put in jail.

Ronald Ein: If we all stay silent, then what we get is tyranny. We will not stand aside.

Betsey Simonds: PASNY lied under oath ... would you want to raise your family 300 feet from this power line? (Judge: "I certainly would not..")

Reverend Robert Simonds: I am a loyal citizen. I have upheld the law and urged others to do the same ... our right to disagree has been violated by this court.

June Black: Some day soon, I swear to God, we will find a way to change these laws so that my children, and all children, will have a future.

Georgette Lauzon: You can do with me what you want, but I shall have to stand.

CHAPTER 13

The Quality of Justice

When I stepped down from the witness stand on 24 March 1977—my thirteenth full day of cross-examination in both hearings—it had been almost three years since Simpson had visited our lab and got Becker and me involved for "one day's testimony." Finally, it was decision time, and the problem was now in the hands of the Commission.

What the Commission would do with its hot potato remained to be seen. The hearings had removed any doubt whatsoever, it seemed to me, about the existence of health risks from exposure to NIEMR, and showed that the power companies' claims for safety had no scientific basis. More than that, the hearings had raised an ever-expanding series of related questions that had to be solved. How *bad* is the risk? How *much* exposure is bad? How much will it cost to eliminate the risk? And what about all the high-voltage lines already operating—what's to be done with them?

The Commissioners had to do justice. I didn't envy them. The Commission had seven members, appointed by the Governor and confirmed by the New York Senate. Since I had become involved, Alfred Kahn had left as Chairman and his job had been taken by a young lawyer, Edward Berlin. The other six seats were apportioned according to standard New York politics: various commissioners were principally identified with a dominant constituency such as environmentalists, women, minorities, industry. The seven-member Commission was supposed to supervise the entire Department of Public Service, which had several divisions. Matias worked in the judges' division; Simpson, in the legal division.

In theory, if the power companies made claims about projects or rate requests that were disputable in the view of PSC's engineers or economists, then the two sides were supposed to go to "court" to argue the points. The engineers or economists would be represented by Simpson or one of his colleagues, and the court would be run by Matias or one of his. Then the evidence from this "trial" would be given to the Commission itself to decide who was right. PSC, like hundreds of similar agencies at both the state and Federal levels, is a uniquely American legal creature for which there is no strong historical model, and it's often hard to figure out who's really in charge because each part of the beast has some authority.

The judge, for example, runs the hearing and decides what will or will not be "evidence." If he denies admission of an item, that's pretty much the end of it; appeals to the Commission are quite rare. By law, the Commission must make its decision on the basis of the evidence accepted by the judge, so that the judge has great power in limiting what the Commission does. Yet the judge works for the Commission, and although they can't fire him—because of civil service rules—they can certainly reassign him to another job.

A judge becomes involved in a particular question only when some PSC engineer or (in rate cases) economist decides there's a problem. In the case of powerline NIEMR, one solitary engineer had made the judgment that RG&E's performance in early 1974 was less than the whole truth. Had he not done that, there would have been no hearings whatever over the health issue. Not only was the engineer not required to make his point, but the system actually worked to impede him from doing so; moreover, in 1974 his suspicions were utterly novel. But the engineer did seek legal help from PSC, and Simpson was assigned.

It is axiomatic among lawyers working for regulatory agencies that they do not anger those in the regulated industry because such lawyers often wind up working *for* that industry. Thus, there was nothing to encourage Simpson to be zealous in pressing his case, and absolutely no one would have complained had he been less than diligent in protecting the public interest. Moreover, unlike Matias, Simpson served at the pleasure of his boss, without civil-service protection. His job situation thus did not encourage bold and innovative steps, especially steps that might cause political fallout. Given these factors, the job that Simpson did in this case still amazes me.

Sitting at the top of this heap was the Commission itself, some of whose members were good and some bad. Some were under Carey's thumb because their five-year term was about to expire and they wanted reappointment. A small minority seemed disposed to make decisions on the basis of the evidence, regardless of the political consequences. Some members were bright and could think incisively and write well; others seemed barely capable. From this political hodgepodge, a just decision was supposed to emerge. I had learned enough about PSC by now to feel very uncertain indeed about the final outcome.

*

Soon after my role in the hearing ended, something happened to exacerbate that uncertainty. Simpson called me one day and said that Francis Wallace, the PASNY attorney, had accused me on the record of possible perjury. Now it is a lawyer's job to try to show that his adversary's views are wrong, but perjury was another matter altogether.

Wallace simply stood up one day when I wasn't there, made the charge, and sat down. He refused to tell Simpson, who was present, any details. I knew of course that Wallace despised me—principally, I thought, because of the things I'd said about him in the letter to Carey (his lack of attendance). But I had never imagined that he would stoop so low. He well knew that such a rap could be utterly fatal to my career as a scientist and my career as a lawyer.

I told Simpson I wanted to know *exactly* what Wallace was talking about—which part of my testimony he was attacking—and that I wanted a PSC inquiry so I could respond to the charge. As things then stood, Wallace remained active in the hearing and hence spoke for the record which was then quoted and used by the power-company PR men in the companies' press releases. I, on the other hand, was out of the hearing entirely and thus had no way to defend myself.

Over the next several weeks, Simpson tried to get to the bottom of the matter. But he was powerless to do anything unless either his immediate boss, Rheingold, decided to do something, or unless Matias himself took some step to investigate the charge. In the beginning, I thought that the latter possibility was more likely because Matias was no fonder of me than Wallace was. I thought Matias might figure this would be a good way to get me if he thought the charges were true. But as the weeks wore on Matias did nothing, and I began to suspect that he *knew* the charges were false and that it suited his purposes to leave me in limbo.

I told Rheingold the thing had to be cleared up, and quickly; he said he agreed with me but that he had been scheduled to take a month's vacation and that if he didn't take it then, he would lose the time. "As soon as I come back, I'll investigate the matter, Andy," he said. I thought the situation was a little more urgent than that, so I wrote directly to the Chairman of the Commission, Edward Berlin, and told him so. Berlin replied that the situation was serious but that Rheingold was the man for the job, and that it would be best to wait for him to return from vacation.

In the meantime, toward the end of May, Wallace repeated his accusation that, possibly, I was a perjuror. This time, however, under Simpson's prodding, he gave a brief explanation of the basis for his charge. He said that when he cross-examined me in March, he asked me whether I had any communication with the landowners in northern New York who were resisting PASNY's line, and I told him no. Wallace said that one of PASNY's PR men who had attended a Fort Covington rally in October 1976 said that a speaker at the rally said she had received a telegram from me which said, "Sorry I can't be there with you today, but I am with you 100%." I'd never sent such a telegram to anybody. The whole thing was totally fabricated.

I was furious that Berlin, Rheingold, or Matias wouldn't do anything, and Simpson too felt frustrated. Rheingold in particular continued to promise me an investigation during which I could tell my side of the story—that Wallace had sucked the entire incident out of his thumb. But Rheingold never gave me the chance. Sometime around June, he quit PSC and went to work for the American Electric Power Company as their General Counsel.

After that, Berlin or somebody else at PSC worked out a deal with Wallace, because Wallace announced in early June, on the record, that he was withdrawing his charge. A little later Berlin sent me a letter saying that the matter was now over, and that I could be "assured that the record as it now stands in no way reflects adversely on your personal integrity or professional reputation."

Fine, I thought, that's great ... he tells me that he doesn't think I'm a perjuror but I'm on the defensive for three months, with my whole testimony—my reputation—at stake, and the man who raised the issue in the first place comes out of it unscathed and unpunished. At this point, I really began to wonder who at PSC were wearing the white hats, and who were wearing the black. This feeling became particularly vivid soon after when Berlin also quit PSC and was hired by Rheingold to defend a 765,000-volt powerline on the basis of testimony given by Sol Michaelson.

*

79

Just about the time that Berlin was telling me he thought I was okay, things took a very, very bad turn. The Commission told *Matias* to make the decision; they said they would review it and make whatever changes they thought necessary, and then make it official.

Now from a practical point of view the Commission's actions made a lot of sense. Matias's work product would be a kind of lightning rod for industry and environmentalist criticisms, and that process would provide the Commission some "feel" for how far to go in one direction or the other when *it* spoke. But for me, the news that Matias was going to write a "recommended decision" was horrible. Matias always seemed to me to have no knowledge or even interest in the scientific issues in the hearing. Indeed, he frequently sent Colbeth in his place. I thought that he held an important job but was bored with it and hence did it superficially, and that the losers were the public. Toward the end of the hearing I made no attempt to hide my contempt for him, nor did he for me.

The idea that he would now be my judge was revolting. I had given almost 3000 pages of testimony in both hearings, and there wasn't a single line in it I would have changed. Also, there was essentially nothing I wanted to say that I hadn't said. A witness simply can't do any better than that. Those 3000 pages were now about to be filtered through the mind of a man who I was sure wanted my hide.

Another thing that bothered me was that Simpson began to put some distance between us. It was inevitable, I guess, and I should have recognized that, but it still hurt. Simpson defended me at the hearing, but he wasn't my lawyer, he was the staff's lawyer—he actually represented the engineer whose dissatisfaction with RG&E in 1974 had started the whole hearing process, and he represented his boss. Now that it was decision time, Simpson was going to write a legal brief and take a position on what ought to be done, and the position he would be defending was not mine but the engineer's. The engineer's position, though ultraradical compared with that of the power companies, was conservative compared with mine because we were all operating from different concerns: the companies were looking at dollars, I was looking at science, and the engineer was concerned with practicality.

So if I wanted to get in my two cents' worth, I had to write my own legal brief and explain why, on the basis of the evidence, the restrictions I recommended to protect public health would be better and more appropriate than those argued for by Simpson. Under the PSC rules I would not be permitted to do so unless I first submitted my brief to Matias. Thus, I was put in the bizarre position of not only being judged by Matias, but of having to argue before him for the mere right to present a brief.

*

In August, Philip Handler finally released the much-awaited NAS Sanguine report. It contained only one surprise, which was more funny than important. The language and analysis in the report tracked almost perfectly with the testimony given by Miller and Michaelson in New York. And, of course, it led to the conclusion that there was no possible health hazard associated with Sanguine.

There was only one bona fide NIEMR investigator on the Sanguine committee: Ross Adey. He published papers that reported biological effects, but I think he managed to survive by almost never commenting publicly on the public-health significance of his work; by walking this delicate line he managed to remain funded while still publishing papers revealing NIEMR-induced biological effects. For many years he has been the only U.S. investigator able to accomplish such a feat. Adey apparently objected to Miller and Michaelson's condemnation of *all* NIEMR investigators, because that would obviously include him as well. This objection resulted in a deal being worked out in which Adey was permitted to write a chapter on his own work for inclusion in the final report. Schwan, in turn, felt that he ought to have a chapter also. The core of the final report thus consisted of Miller and Michaelson's hatchet-work, and two addenda by Schwan and Adey. This was then "adopted" by the other committee members, who were professed or de facto nonexperts, and the final thing was issued under the imprimatur of the National Academy of Sciences.

Immediately after the report was released, Harvey asked Matias to enter it into the New York record. Matias re-opened the record, entered the report into "evidence," and then closed the record again, all in his office and without a chance for Simpson or me to do a thing about it. Matias said the report was important because it would show what a body of "independent scientists" had to say about me.

My original fears about the NAS committee and what it might do to me were thus confirmed. I had felt all along that Handler, who had the sole authority to decide *when* the report would be released, had been purposely delaying it until the New York hearing was over. Initially it had been scheduled for release in early 1977, but as the hearing dragged on and on, the release date of the report receded. This postponement was clever because it tended to protect the report from controversy; had it been released during the hearing, Simpson would have laid out on the record the history of the committee and probably destroyed its credibility in the hearing. He might even have succeeded in subpoenaing Handler and Hastings to testify. But with the hearing over, a direct attack on the record wasn't possible. The only course open to me now was to attack the NAS report in my brief. Clearly if Matias pointed to the report in his decision, it would damage me; his decision was bound to be widely publicized.

*

Although I was worried about what Matias would do, I also knew he had a tremendously difficult task. He had to write a decision that made sense. There were thousands and thousands of pages of testimony; he could point to any part of it, indicate that he believed or accepted it, and base a decision about a certain point on that. But he had to point to *something*, some evidence, and it had to be related to the conclusion he drew. I didn't think he had the ability to do that. His handling of the hearing had been so bad, and his understanding of the issues so superficial, that I couldn't see how his decision could be anything but scientific doubletalk.

However, some time in the fall of 1977 Matias found a way to solve his problem.

81

At a meeting of a NIEMR committee, Cyril Comar of EPRI let it slip that Matias had hired a California physicist to write the decision for him. The man's name was Asher Sheppard. I tried to check out the rumor with PSC, but their official position was that nobody in any of the divisions knew anything about Sheppard or about his being hired to write Matias's decision. The judges' division was mum—they weren't saying *anything*.

Sheppard was a young man who had recently finished his Ph.D. studies in the East, and then had moved west to work for Ross Adey. When I first met him, in 1975, I had liked him. He had come to visit me in the lab for a day, said he was impressed with my work, and then had written later to say that my PSC testimony was "interesting and well-presented and, in many ways, it presents views similar to mine." But after Sheppard moved to California my opinion changed. First, I heard he had become a power-industry consultant on NIEMR-related problems. To me, that alone—measured against his earlier comments to me—was enough to put his scientific objectivity in doubt. Then I heard Sheppard had apparently become a proponent of Adey's school of thought about NIEMR: "Let's keep quiet until we know everything there is to know." In this view, the public is seen as too dumb to be told the truth about NIEMR problems because they won't be able to deal with all the uncertainty and anxiety created. The truth might lead to panic, so it would be better for us scientists to keep all this information to specialist journals until we had all the answers; then, in one move, the public could be informed both that a problem existed and that the solution to it had been worked out. That way industry wouldn't have to risk erring on the side of safety, i.e., spending unnecessary money.

But what really settled the issue of Sheppard for me was a review he wrote of one of my articles. I had sent it to a journal for publication and all the reviews that came back except one were quite favorable. The reviewer who didn't like it said our work was of "poor quality," and that the studies we had described were not "accepted by the scientific community." He said that "one need not be an alarmist at the hint of a biological effect." Finally, he said my paper was an "anti-intellectual attack," that it had a "nasty tone," and that I had taken a "cheap shot at the electric utilities."

As is the custom, the reviews were anonymous. But this one sounded as if it had been written by Adey or someone with similar views, because the author's hidden agenda was clearly that there *were* biological effects of NIEMR, but only a wild-eyed radical would talk about them before all the evidence was in (which might of course be decades). The review also contained comments about the physics in my article, and I knew that Adey was a physician, not a physicist. Adey had only one physicist on his staff—Asher Sheppard.

To ascertain whether Sheppard had actually written the review, I sent him a letter in which I raised some innocuous point that had come out of the flow of our correspondence over the previous two years, and toward the end I wrote, "By the way, Asher, in that recent review of my article, you wrote that…" and I went on to dispute some trivial point about a comment involving physics which had been made in the review. Sure enough, Sheppard took the bait. He wrote back to clarify the point I had raised, and in the process confirmed that he had written the review. That was all I needed to know to make sure that Sheppard had joined the opposition. There was no doubt in my mind that in his anonymous work on Matias's decision—writing *as* Matias—he would let me have it with both barrels.

*

Matias's decision came out in January 1978. It was signed by him and Colbeth, and Sheppard's role in it was not acknowledged in any way. It attacked me from practically every direction.

One of Matias's main props was the testimony of Henry Hess, the power-company computer man who had statistically "analyzed" my data. Matias quoted Hess at great length and concluded that his testimony showed my experiments to be worthless and "unsuited to the formation of reliable conclusions." During the hearing Harvey had asked me what textbook I had used in studying statistics in graduate school; Matias (or more likely Sheppard) found the book in a library and read that the author intended it to "introduce the undergraduate student to unifying concepts of probability in statistics." Since I had used a book intended for undergraduates, Matias said, that proved that I was "not an expert in the field of statistics nor in the ancillary area of the design of experiments."

Throughout the decision Matias drew conclusions like this one that went far beyond his flimsy "evidence." Quoting Hess repeatedly to the effect that I was an exceedingly poor scientist and that my work therefore could not be believed, Matias concluded: "Based upon the clear and admitted lack of expertise in the use of statistical tools to design experiments and to analyze the measurements obtained therefrom by Dr. Marino, and from the testimony and exhibits sponsored and prepared by witnesses Marino and Hess, it becomes intuitively obvious that there is no substantial evidence in this proceeding which would lead to the conclusion that there is any reasonable probability that 60-Hertz [NIEMR]…would cause any significant [biological] change."

I might have thought that that was enough, even for Matias, in terms of undercutting me personally. I mean he could have figured that people would believe either him and Hess, or me, and there was probably no need to continue attacking me because that might be counterproductive. But he *did* go on to try and get me personally—not as someone who does scientific experiments but as a person. To someone looking at it from outside, Matias's decision must have seemed seriously unbalanced.

The first thing he did was criticize me for insisting that the hearings that I was involved in be held in Syracuse. He said that that caused "a degree of resentment" on the part of the other participants in the hearing because of the "inconvenience" it caused them. The impression that this section of his decision conveyed was that I was a thoughtless, uncaring person who selfishly inconvenienced many other people.

Matias discussed at considerable length the letter I had sent to Governor Carey. He said that there were "some very serious misstatements of facts" in it, but he never said exactly what they were. Matias sort of suggested that one of them was my statement to the Governor that the lawyers for PASNY didn't always attend the hearing. But he never said explicitly that it was a misstatement of fact, because that would have been untrue—he simply *implied* it. One statement in the letter that Matias called "misleading" was my telling

Carey that I was being cross-examined by a nine-lawyer team. I hadn't explained to Carey, Matias said, that although nine lawyers were part of the team, no more than four or five were usually in the courtroom at any one time. The gist of his analysis of my letter to Carey was that I was among that low species of animal that would intentionally attempt to deceive the governor of a state.

Matias went into great detail about the fact that I had been interviewed on "60 Minutes." He lauded Michaelson and Miller for not appearing because he said the "issues" were being litigated in his courtroom and it was therefore improper to talk about them publicly, "particularly when sensationalism may attach to those issues." Matias said he found my appearance on the show "inexplicable." In regard to the report that I had written for "60 Minutes," Matias implied that I had hidden it, or refused to make it available, because it probably contained stuff that would hurt me. From all this "evidence" the reader was meant to conclude that I was a publicity hound and uncooperative, and tried to hide things.

Matias also quoted at length from the letter that the Johns Hopkins engineer had sent me regarding the actual NIEMR strength I had used in my experiments. He couldn't use the letter to show that the NIEMR strength I had calculated was wrong, because even the power companies conceded that the engineer was wrong. So Matias used it as "proof" that other scientists disagreed with me. And he made this point even more forcefully in the context of the NAS Sanguine committee report. He quoted from what he called a panel of "distinguished, nationally known scientists" to show that these distinguished scientists disagreed with me.

Matias frequently laced his remarks about my character with attacks on my scientific competence. The result was enormously heavyhanded. For example, after making the point that the NAS experts disagreed with me, he accused me of "highly questionable" scientific practices and said that "Dr. Marino's experimental work was not conducted carefully enough for the results to be believable."

Matias barely mentioned the forty-five studies that I had cited as evidence that there would be a health risk from powerline NIEMR. Instead, he bought Harvey's argument that I was a bad scientist and a bad person, and that the effect of this characterization was "to diminish trust in his judgment of the literature he has cited." Harvey couldn't have hoped for better language because the position that Matias took was that *I* was the foundation for all the studies—that my own personal qualities and characteristics governed their credibility. Thus, if I was a liar or incompetent, the studies could have no support in the record.

Matias's overall conclusion regarding me was as follows:

In this recommended decision it has been demonstrated that Dr. Marino was reckless and inaccurate in his public statements, that he exaggerated in his letter to Governor Carey, that he was evasive and argumentative under cross-examination, and that he was unable to cite any support by other scientists for his speculation as to the biological effects from powerline [NIEMR]. With this background the indications of careless procedures, faulty statistics, and unsupported extrapolations from his own experiments make it impossible to place any substantial reliance on Dr. Marino's scientific research set forth in this record.

One of the greatest aberrations of Matias's decision was that it practically ignored Becker. He was dismissed on the grounds that his testimony was based on my testimony (Harvey's "me too" theory) and that since my testimony was worthless, Becker's couldn't stand.

After all these shenanigans you'd expect Matias to conclude that powerline NIEMR was not a health risk, right? Wrong!

He concluded that the lines *were* a health risk and that preventive steps had to be taken to protect the public against involuntary exposure. He said: "What is necessary is to remove the involuntary feature, that is, to insure that persons living or working near the line are not involuntarily exposed to danger and that persons who enter the right-of-way do so voluntarily with knowledge that chronic long-term exposure may entail some risk." He also said: "The one solid conclusion that can be drawn from all scientific testimony in this case is that there are NIEMR interactions with biological organisms which cannot be adequately explained on the basis of current knowledge." Then he concluded: "Thus we find that transmission lines should be built and operated so that no person works daily or lives in a [NIEMR] field greater than 1 kilovolt/meter."

The structure of the recommended decision made absolutely no sense. On the one hand, I was the only witness to testify to the existence of all the studies—a fact that Matias readily conceded—and, as a scientist and a person, I was called everything but decent. How then was it possible for Matias to accept my conclusions? Harvey and Wallace were driven crazy by the decision, because the "analysis" section had gone their way but the conclusion had come out of thin air. The two lawyers complained bitterly to the Commission about the illogicality of the decision, and several individuals tried to get an explanation from Matias and Colbeth, but they adamantly refused comment on what they had done and why.

I simply didn't know whether to laugh or cry. On the one hand, the attacks on me hurt badly—I felt as if someone had rolled the wheel of a car on my chest. A lot of attention was given to the recommended decision in the press; the day it came out, the gist of the news story on local TV was that the chief witness against the powerline had been found "reckless, evasive and argumentative" and that his "research was not believable." As the reporter did the story, such quotes flashed on and off the screen as subtitles for the report, and each flash burned my eyes. The decision was reported in the *New York Times* under the headline "PSC Judges Find No Peril in Controversial Power Line." Later the *Times* editorialized on the safety of the line, emphasizing what Matias and Colbeth had said about my recklessness and unreliability.

On the other hand, Matias had accepted two of my three major points: (1) a health risk *did exist* because there were bona fide scientific studies that showed that animals (and, in some cases, humans) were affected; and (2) it was wrong to expose people to powerline NIEMR involuntarily.

The third point I had urged throughout the hearing, a point that Matias rejected, was the creation of what I called an Administrative Research Council (ARC). I wanted a group of independent scientists assembled and authorized to fund experiments by independent scientific investigators. The only condition for obtaining money from ARC would be that the proposed experiment be done by a competent expert, and be designed to provide data about possible health risks. There would be no tacit agreement—as was usually the case with power-company money—that the investigator find nothing contrary to power-company interests. This suggestion Matias rejected flatly; research into the possible health risks of powerlines, he said, should be done by the companies or some Federal agency.

After my initial personal distress, I began to understand that Matias's schizophrenic decision reflected, first, his recognition that there was a problem with powerline NIEMR and that something had to be done; and second, his wish to neutralize the person who had raised the problem, me, and thus prevent me from marching through the various public service commissions around the country and causing problems for the power industry in general.

But regardless of his view of what was at stake, it seemed to me that his having hired a power-company consultant to write his decision for him was a dangerous risk which, if it were found out, could seriously compromise his bosses at the commission, and possibly lead not only to public scandal but a lawsuit. In the final analysis, and aside from his dislike of me, I think Marias was motivated principally by desperation. He simply didn't know what else to do.

CHAPTER 14

Ways of Deciding

In the dog days following Matias' recommended decision, as we waited for the Commission to make up its mind, I had time to reflect on the meaning of the experience I'd gone through. What especially interested me was its relevance to the general problem of how to resolve technological disputes.

As a lawyer and a scientist I was really divided about the PSC procedure. I had a pretty good notion that had I not been a lawyer, we might never have survived all the maneuvering. As a scientist, on the other hand, I thought it unfortunate that the rules of evidence rather than scientific rules were followed: this procedure created an enormously complicated and often confused hearing record. Also, the PSC hearing had allowed the political motives of the participants to be obscured and confounded with the science in a way the public could not perceive. Furthermore, the final decision itself was bound to be deeply affected by politics, especially the response of the various members of the Commission to their respective constituencies, and that was also something the public did not understand. Finally, whether the public could have faith in the justice of the final decision was extremely uncertain, especially in view of the general perception in the North Country that PSC and PASNY were in bed together.

On the other hand, I knew of no precedent for the role Becker and I had been able to play in the process. We had pressed our views for three years, our words had been recorded, and the Commission *said* they would be taken fully into account. We had been defended by a dedicated lawyer, Bob Simpson, after the initial decision by a forthright engineer that a problem did exist and should be examined. So we had to admit that while there were people like Matias, there were also good people in authority who took their jobs seriously.

Also, because of the extensive press coverage the public knew at least that a serious debate was going on. That was very important. I had recently acquired several thousand pages of previously secret documents concerning the "Moscow signal" affair—the irradiation of the U.S. Embassy with microwaves by the Soviets—and as I perused them it was clear to me that letting the Federal government deal with such issues secretly was the *worst possible* way of proceeding. I felt the documents revealed bad faith, bungled and hidden studies, and attempts to keep the Embassy employees themselves in the dark while the problem was studied. Moreover, the alleged national-security dimension of the issue had assured that the public would not know what was going on. In addition, these papers did not indicate that any real decision had been made about the principal issue—whether the personnel were being harmed. Clearly, this was no way to resolve the question of NIEMR health risks. So you had on the one hand the long and messy and acrimonious but public debate within the PSC hearing, and on the other the entirely secret "deliberations" of the U.S. Departments of State and of Defense in the Moscow affair. There had to be some better way. [See Paul Brodeur, *The Zapping of America* (Norton, 1978), and Nicholas Steneck, *The Microwave Debate* (MIT Press, 1984), for accounts of the "Moscow signal" affair.]

Of course for some years that "better way" had been the choosing of blue-ribbon panels to settle disputes, where presumed experts debated the issues among themselves and then issued a report. But this had been the NAS way, and my experience with that had been horrendous. The Navy had asked Philip Handler to choose a committee to settle the Sanguine/Seafarer issue, and he had made very bad choices. The NAS Sanguine panel was simply a group picked to rubberstamp a predetermined decision. The most crucial point with blue-ribbon panels was that *someone* had to choose them; the NAS panel could only be as good as Handler's wisdom and fairness in choosing them. And in addition there had been secrecy surrounding the panel's deliberations because NAS was not subject to the Freedom of Information laws. The panel went into a back room, did its work, and issued the report, and the public knew nothing about the process.

I wasn't the first person to come to such conclusions about NAS. Handler had been embroiled in several similar controversies involving stacked committees. [See Philip Boffey's book, *The Brain Bank of America: An Inquiry into the Politics of Science* (McGraw-Hill, 1975).] In fact the growing concern over this situation was one factor that led to proposals in the early 1970s to establish what are called "science courts" to resolve controversial issues.

The idea of the science court was that, first, the adversarial nature of the specific dispute would be openly admitted—rather than partly masked as in the PSC procedure, or totally masked as in the Moscow signal and the Sanguine affairs; and second, the values in which the dispute was embedded would be separated from the scientific questions, the matters of fact. It would be the scientific questions alone that would be decided by the science court. Each side in the dispute would draw up clearly defined statements of scientific fact—the process to be conducted by a referee chosen by both sides—and the case would be tried before a panel of scientist-judges who had also been agreed to by both sides. The expertise of these scientist-judges would be in areas external to the areas of the dispute.

A Presidential task force had been set up to explore this idea; one of its members was Allan Mazur, a professor of sociology at Syracuse University whom Becker and I knew. In the midst of the hearing Mazur had approached us with the idea of testing the science court idea on the powerline issue. Mazur wanted to get each side to submit their statements of scientific fact to the other, with the aim of trying to arrive at gradually refined statements on each side to which the other side could subscribe; or if that didn't work, to take the dispute to a science court. The two sides would jointly select the judges, present their evidence, and cross-examine each other. Then the judges would make their decision, specifying the reasons on which it was based. The entire procedure would be open to the public.

That sounded like a great idea to me. In 1976 Mazur sent letters to everyone he thought might be interested in defending the safety of the powerline, explaining to them the idea of the science court and enclosing an article the task force had written for Science. In his letter Mazur also enclosed Becker's and my statements of facts about biological effects of NIEMR, and asked the proponents of the line to respond with theirs.

Thomas Matias, Public Service Commission hearing officer, listening as the late Glenn Bullock, an area farmer, speaks in public hearing against powerline during Winter 1976. (Photo by Doug Jones.)

Largest demonstration against powerline: 1000 people marching through Edwards, N.Y., in March 1977. (Photo by Doug Jones.)

Portion of power company legal team at hearing in Syracuse, N.Y. Robert Harvey is in center (arms folded), Francis Wallace is to his left. (Courtesy Watertown Daily Times.*)*

Robert O. Becker, M.D.

Completed 765 000-volt powerline tower: concrete footing at base of tower is approximately 3 ft high. (Courtesy Watertown Daily Times.)

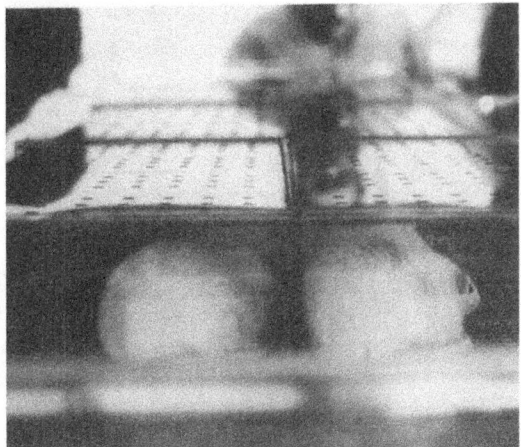

Rats used in one of the studies conducted by Richard Phillips, Battelle Northwest Laboratories, Richland, Wash. Cages are 4 in. high.

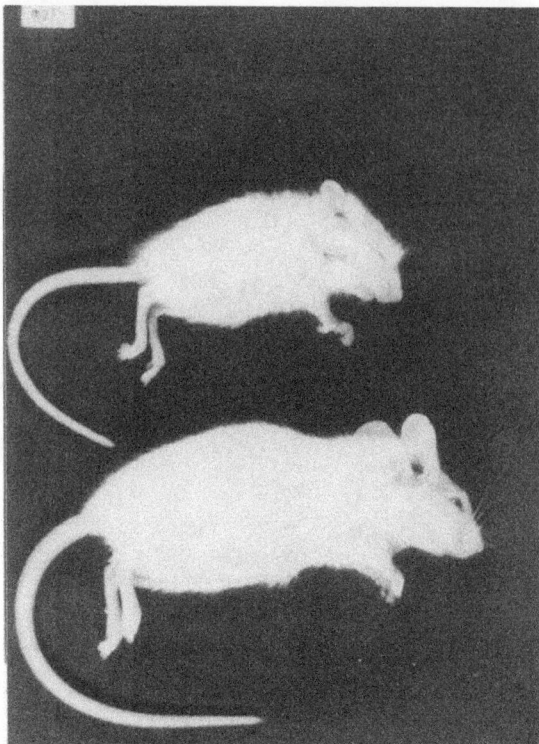

Female mice from Marino's 1976 three-generation NIEMR study: bottom, normal control mouse; top, mouse continuously exposed to NIEMR from conception. Both mice are approximately 100 days old. This is the picture of the "woefully stunted mouse" to which Handler referred (p. 100).

NIEMR from high-voltage powerline will cause ordinary 40-watt fluorescent bulb to glow.

The response wasn't entirely unexpected. Even after I had refined our list of facts to make them more specific, the proponents were unwilling to respond with their own lists of facts to the contrary.

But what was most revealing was that none of them even wanted to *participate* in a science court. Michaelson refused because he said a science court would merely give more publicity to views that were wrong. Miller refused because he said science was nonadversarial; a thing was either right or wrong and there was an end of it—nothing to argue about; besides, he said, he had no time. Schwan initially seemed interested, then begged off. Mazur was unable to find a single person opposing us who was willing to take part.

The whole affair took many months to complete. What we eventually concluded was that the side which perceived itself as already having the advantage in a dispute would see the science court as a threat. The proponents of the line still perceived, I think, that they were in control of the issue because Becker and I were two outsiders who had the burden of proof. They could rely on the general public ignorance of problems with powerline NIEMR. A science court proceeding on the issue, which would provide Becker and me with more publicity and thus legitimize our views in the minds of the public, would undermine that ignorance and give people ideas. The powerline proponents apparently felt that merely going through the process would be a victory for us.

Of course enough acrimony had developed in the hearing that such a proposal to our adversaries was probably bound to fail. But even though our experience was negative, that didn't convince me that the idea was unsound. There were, certainly, difficulties with it. How would a science court have enough standing to encourage the proponents of a disputed technology to join in? And how could it gain the authority to force disclosure of information? Yet it seemed to me that it was in theory the closest thing yet proposed to the *possibility* of unbiased and open decisionmaking.

CHAPTER 15

Final Decision

With the issuance of his recommended decision in January 1978 Matias was through with the hearing, and the ball was now in the Commission's court.

They must have been extremely irritated with Matias's decision, because it shot to hell all their perfectly legitimate reasons for directing him to write a decision. For one thing, the technical analysis of the record made no sense. It was obvious throughout the hearing that one side or the other would appeal the Commission's final order to the courts; but when and if that happened, Matias's decision would never have passed muster. The second thing was that Matias had made very few practical suggestions, which the Commission certainly had wanted him to do. And third, the commission must have been very worried about the corruption of the decisional process itself because of Matias's reputed hiring of someone else to write the decision. Surely that would have come out when the whole process was reviewed by the courts—and it would have reflected terribly on the Commission itself, which was supposed to be supervising Matias.

The Commission, then, had no choice but to undertake the dirty job of reviewing the huge record itself. It took possession of all the documents and gave them to a lawyer with instructions to draft a technically correct analysis based on the record.

But the lawyer needed to know what the conclusion was supposed to be so that his analysis could be shaped accordingly. The Commission mulled this problem over in March and April, in private. I knew damn well what was going on, and it certainly was not what was supposed to happen, or what the public thought was happening. The individual commissioners were making up their minds about what to do not so much on the basis of the record as on information from sources beyond the hearing: newspapers, magazines, TV and radio reports, and private conversations with all sorts of people. All this data input, of course, was filtered through their own personal prejudices and predispositions and ambitions. I would have bet a dollar to a doornail that Governor Carey had also let it be known what he thought should or should not be done. What actually happened was that the individual commissioners made up their minds, and then a kind of consensus had to be developed among them. Part of this process took place in public, and it tended to augur good things as far as I was concerned and the views that I held.

In early May, for example, the Commission unanimously agreed that their final decision would not contain any attacks on my character or ability as a scientist. They specifically rejected all Matias's fulminations against me. That was very important. If they didn't want to believe what I said, well, that was their prerogative; but to call me a string of dirty names would be just plain overkill. Matias's language had really hurt me, upset me; I was afraid of its implications for my career, both as a lawyer and as a scientist. I envisioned journal editors who were reviewing my papers finding out about it ... people who reviewed my grant applications would know of it ... would *they* think that the work I did wasn't worth anything? What would a judge before whom I was arguing think? What would a jury before whom I was testifying as an expert witness think of Matias's language? Surely the lawyer for the other side would make sure they heard it.

The Commission's decision on this point took a lot of pressure off me. But it was good for them, too, because nobody can throw mud without getting it on themselves; even if the Commission hated my guts and disbelieved me entirely, it would be better for them to reject my testimony in measured, unemotional terms that conveyed a sense of logic and professionalism. And it was good because if they had concurred with Matias, they would have succeeded in scaring away anybody willing to testify in the public interest as an expert witness before the commission. Certainly from my experience, all a witness can expect is grief and abuse; at the worst, he may run into another Matias. The very least an expert should get from voluntarily participating in such a hearing is the knowledge that he won't be abused. So if the Commission had followed Matias's lead, it stood to reason that future witnesses would have to be practically insane to agree to testify, even if they were paid.

Aside from this dimension, the Commission was saddled with a big responsibility in deciding what to do about the NIEMR. The emerging feeling was that there *had* to be a NIEMR health problem, or else the whole thing would have blown over long ago. What was to be done about it? The commission would be writing on a clean slate; it was in an area where it had little expertise; it was subject to great gubernatorial pressure as well as furious attacks from the residents of the North country; and there was pressure from all sides to do *something* quickly.

*

In June 1978 the final decision came out.

The *first* thing I did was to look through the hundred-odd pages to see what it said about me personally. Only after that was settled could I bring myself to read what the Commission said about the NIEMR health risk.

First and foremost, the commission accepted my view that the animal studies showed there was a human health risk. They wrote: "The record before us ... contains unrefuted inferences of possible risks that we cannot ignore. [NIEMR] similar to ... that here at issue seems to have produced effects in laboratory animals, and these effects cannot be presumed harmless." With that simple language the commission took giant steps. It said that there were at least some valid experiments—they weren't all done by incompetents as the power-company witnesses had testified. Second, the Commission said that, even though these were mainly animal experiments, it had to consider the results in the context of human exposure along powerline rights-of-way.

These "findings of fact" were completely unprecedented in the area of NIEMR health risks. As far as powerlines in New York were concerned, and

91

probably those in the rest of the country, they opened Pandora's Box. Now all the other even more difficult questions would have to be considered.

The Commission didn't embrace every experiment that I described, but neither did it buy the power companies' argument that an experiment could not be accepted unless it was free of the *possibility* of error. The Commission said: "Some of these experiments may have been flawed in various ways, but we are far from persuaded that their flaws warrant disregarding their results."

One of the things that had always galled me was the way the power companies treated my experiments. Although they spent an enormous amount of money to analyze them and criticize them, they never made money available to reputable people to duplicate them. The Commission also thought that was pretty strange, and they said: "[The power companies] could have aided the record by attempting to replicate Dr. Marino's experiments in a manner free of the defects they perceived in them. For reasons best known to them, they did not do so." That comment gave me a lot of pleasure.

The boldest decision the Commission made was to accept my idea for the creation of an Administrative Research Council. Once more overruling Matias, they said that more research on the health issue was definitely needed and that they weren't going to depend on the power companies to do it for them. The Commission said that it would create the ARC and authorize it to spend money for a 5- to 7-year research program. This decision, too, was unprecedented. Previously, virtually all the money available for powerline NIEMR research emanated from industry (or industry-controlled) sources, or the Federal government, and a truly independent investigator had as much chance of being funded as of going to the moon.

In a sense, the Commission's decisions were courageous; no regulatory agency had ever concluded such things about powerlines. But in another sense they were easy because they amounted to mere words and promises. The hard part was action. Even here I think they came through. The PASNY line was already built and operating—that was simply New York politics—and there was absolutely nothing the Commission could do about that. However, it had held onto the right to "fine-tune" the construction of the line to accommodate the conclusions that they reached in their final decision. Working within this framework, the Commission decided that the right-of-way would be increased by 40% in order to insure that people would be kept farther away from the line and hence in a region where the NIEMR was somewhat less intense. In addition, the Commission ordered that residents living within a zone of 1200 feet should be given the option of having their homes relocated or purchased by PASNY, so that no one would have to live so near the line unless they wanted to.

The RG&E line was something entirely different. Beginning with the Commission's final decision, RG&E and NiMo's problems mounted precipitously; ultimately, they were denied the right to build. It had taken more than five years, but the Rochester people who had initiated the battle against powerline NIEMR finally had what they wanted, which was to be left alone. Then, not long after the final decision was handed down by the Commission, PASNY itself announced that because of the outrage the line had caused, it would build no more 765,000-volt powerlines. If PASNY can't build them in New York, they can't be built in New York.

The final decision contained one major disappointment for me—the point about human experimentation, and warning people. Despite the furor that surrounded the hearings in New York, there were still many people who had no idea that not only the 765,000-volt powerline but *all* existing high-voltage powerlines ate surrounded by NIEMR which, when one is constantly exposed to it, can create a health risk. There were many miles of such lines—230,000- and 345,000-volt lines, for example—in New York. I thought it was incumbent on the Commission to some kind of a notice or warning to everybody who lived near them. I knew it wouldn't be the easiest thing in the world to draft such a warning, but I also knew it could be done, and I was absolutely convinced that it had to be done. These people had to be told, at the very least, that there was a bona fide scientific dispute about possible risks, and that the final word was not yet in.

I figured that there were some people who preferred to raise their families and live their lives while erring on the side of caution. For example, I don't let my kids eat cereal with BHT in it, not because I know that it's harmful to their health but because I know that the tests that have been done are inadequate and therefore that it *might* be bad. If people knew about the possibility of a problem with powerlines, they might want to move, and not run the risk. They *had* to be given that option, I felt. Even Matias had said that it was necessary to "remove the involuntary aspect" of exposure. Imagine that—even Matias. But the Commission must have been afraid there would be a panic if it issued such a warning notice, so it didn't.

The net result was the following epic irony: It would be unethical and illegal for a highly competent professional scientist to expose a person to powerline NIEMR in a laboratory experiment without first obtaining his informed consent, regardless of how well the situation was monitored and controlled, and regardless of how many precautions had been taken to insure that the subject would not be injured. The investigator could have an army of physicians and an ambulance waiting at the front door in case a problem developed, but it would still be unethical and illegal unless he first *asked* the subject for permission. Yet the local power company could expose a citizen to virtually any level of NIEMR without securing permission, without telling him it was being created in his home or on his land, and without monitoring or checking the health consequences in any way whatever.

That's the way it ended ... or began, depending on one's point of view. RG&E and NiMo were terribly disappointed: they came away with absolutely nothing after almost five years of effort. PASNY also was unhappy with the Commission's actions because it hurt them in the only place that such an organization can be hurt—in their pocketbook.

Simpson seemed to think that the end result was worth all he had gone through. His eyes were, much more than mine, fixed on what was practical or possible. He knew his agency pretty well, and he thought it could not have done any more than it did, given the fact that this was the first time it had ever really considered the issue that we presented.

Becker saw the whole five-year effort in much larger terms. After May 1976 he had never been in the trenches again, and I think that helped him see some things more clearly than I could, but also caused him to see other things more simplistically than they actually were. To Becker, the hearing, underneath all the posturing and maneuvering, was actually a conflict between two opposing philosophies of science, both of which had long historical roots. The testimony of Miller, Michaelson, and Hess was simply garbage to be ignored; it had no intellectual value, but was only a series of *ad hominem* attacks. Schwan's testimony, on the other hand, embodied a classic view of the nature of science—it was wrong but at least it was a view, not merely a series of accusations. Schwan tended to emphasize the primacy of what was known in evaluating new information. If one looked at the laws of electricity and magnetism, for example, one could gather no hint that low levels of NIEMR would cause stunted growth in rats or alter the cardiovascular system of dogs. Against this rock-solid body of theoretical physics, the biological observations looked puny, and it was therefore the observations which were to be suspect.

In the light of Becker's own research career, especially, Schwan's view was diametrically opposed to what Becker thought was right. It ignored the overall history of science as well. For example, Becker would point to the case of Galileo, who was forced to recant the validity of his celestial observations because they seemed to conflict with current theological concepts; or to Semmelweis, the Austrian physician who was criticized by his colleagues for persevering in his belief that poor sanitary conditions attendant on childbirth were the reasons for the alarmingly high morbidity and mortality associated with childbirth in nineteenth-century Europe. (The wisdom at that time did not provide any theoretical reason for physicians to wash their hands, so they saw no reason to do so.) There were many other examples but what they all pointed to was the issue of what was going to be regarded as "scientific" when a conflict occurred between orthodoxy and new information.

To Becker, then, the PSC decision was something entirely new in the history of western science: it was the *first time* that the question of orthodoxy vs. new information had been decided by a fair and impartial group of judges in favor of the primacy of new information.

It was hard for me to see the decision in such terms, because it had issued from such a flawed process, and because some of the personal wounds I had suffered from it were still open. Perhaps with time I will come around to Becker's view. But I suppose that partly depends on the sequels. The PSC decision certainly did open the way toward a much-needed recognition that the present pattern of human exposure to NIEMR is not safe. If as a society we ultimately decide not to do anything about it, it may be that I shall come to see the New York hearing as a futile waste. But even if society does make such a decision, it won't be because the evidence for the correct decision doesn't exist.

PART IV

CHAPTER 16

Aftermath

Not long after the final decision, I found out some of the details behind what Matias did. The power companies—especially PASNY—continued to broadcast his attacks on me in the press, despite the Commission's decision, and I was determined to find out the whole story so I could protect myself.

Through my friend Robert Liegel, a Syracuse attorney, I wrote the Commission asking for complete documentation regarding any outside help Matias had had in writing the recommended decision. I wanted everything: copies of agreements he had signed, correspondence, and a copy of Sheppard's final report. Initially I was told they had no such information. I wrote back to say that I *knew* I had a legal right to the material under the Freedom of Information laws, and that if it didn't come, I would sue PSC itself. That's when it came.

It turned out that Matias, working through Colbeth, had contacted Sheppard and hired him for about $5000. Sheppard was given part of the hearing record and told to analyze it and write a report. Colbeth told Sheppard: "We hope to put your report bodily into our decision with a few changes for consistency of style."

What I wanted most was Sheppard's report, but PSC told me that Matias had destroyed it and that no copies remained, either in Matias's personal file or in PSC files. I also wanted to know who okayed the spending of the $5000. I had a very difficult time trying to understand the financial structure of PSC, and they did little to help me along. In the end what I *think* I learned is that the money was authorized by Matias's boss in the judges' division, with no further review at a higher organizational level within PSC. I was told that the judges' division had a certain independence from the rest of PSC. My best guess is that the Commission itself didn't actually know about what was going on at the time.

But after I made my Freedom of Information demands the Commission sure as hell learned about what Matias had done. He had hired someone to do *his* job, and after spending the taxpayers' money to do that he destroyed the report and put on a public face as if he had done all the work. But far worse than that was the violence Matias had done to the legal process itself.

We had a hearing that lasted nearly four years *precisely for the purpose of producing a record*, and the law specifically said that the decision had to be based on that record. But Sheppard's report was based on only part of the record, and much of the material that would have favored my position was never sent him. In addition, Sheppard had read books and articles that were never even *mentioned* in the hearing, and directly quoted them or relied on them in his report. Finally, Matias relied on the statements of Wallace and Harvey about me, and they were neither under oath nor subject to cross-examination. To use lawyers' words as *evidence* is one of the worst things a judge can do.

So Matias hires a power-company consultant to write his decision, pays him $5000, takes the report which contains a significant amount of information from outside the record, puts his name on it, and then destroys Sheppard's original work product. You can guess that this got Matias into a lot of trouble with PSC, right? Wrong. PSC never did anything to him, and today he's still doing the same job. Colbeth died a few years later.

After discovering all this background information I wrote to Sheppard and tried my old bait-and-switch again, but this time it didn't work. I told him about my research, expressed the hope that he and Ross Adey were getting along well, and then mentioned that I knew about his work for New York and asked him for a copy of his report. I said I had heard that Cyril Comar at EPRI had a copy, and that it was only fair that *I* have a copy. Sheppard wrote back that he had talked to Comar, and that Comar didn't have a copy. "Only Tom [Matias] and I have copies," he said, and he wanted it left that way.

*

PASNY sued PSC over the final decision, arguing among other things that PSC was bound by Matias's decision because he was supposed to be the Commission's eyes and ears, and thus that there was no reason in the record to do *anything* in terms of the NIEMR problem. The court flatly rejected the argument and said that PSC could believe anybody they wanted.

Furthermore PASNY argued that *it*, not PSC, had the expertise to evaluate powerline NIEMR health risks, and that PSC ought to keep its hands off the issue. PASNY specifically argued that PSC had no right to demand that the utilities pay for a research program to be run by PSC. The court rejected that argument, too.

But it did buy PASNY's third argument. PASNY didn't like the idea of having to buy the property of people who lived near the line if those people decided within a year or two after the line was built that they wanted out. The court said that whatever restrictions were put on PASNY's right to build the line, they had to be in operation at the time the line was put into service, and couldn't be added on later. This decision ignored the possibility that health effects might turn up well after the line had gone into service.

Neither PASNY nor PSC was satisfied with the court's decision and both sides began preparing for an appeal to the highest court in the state. The spectacle of two state agencies wrangling with one another in the courts did not present a very wholesome public image of the Carey administration, so after more than a year of public bickering and private confabs, it was announced in early 1980 that the case had been settled out of court on the following terms:

There would be a research program, PASNY and the other utilities would pay $5 million for it, and the widened right-of-way would stay in effect. *But ... the research program would be partially run by PASNY.*

*

The ARC, as I had proposed it, was to consist of a group of scientists who would review proposals for research from independent investigators that dealt with evaluation of powerline NIEMR health risks. The only criteria for funding, as I envisioned it, were competence and relevance. My idea was to provide for the existence of an in-house group of scientifically sophisticated people who could recognize what was and was not an appropriate study, and who would make honest judgments. I simply wanted a source of money from which good, unbiased investigators could get funded. What emerged over the next eighteen months was a black beast.

My idea was perverted from the beginning by the PASNY lawsuit, because the settlement required ARC to be run by a troika consisting of PSC, PASNY, and the Health Department. God knows PSC is not the ultimate in bureaucratic efficiency and responsiveness, but I had given it four years of my life and it had responded reasonably well at the end, so I felt some lingering affection for it. Besides, Bob Simpson was to be the PSC staff person on the program. But PASNY, I felt, was beneath contempt and the Health Department was not a whole lot better. Like almost every Health Department in the United States, it had come into being and grown in an era when the major health problems amounted to the control of infectious diseases, and that was still the only thing it really knew how to do well. This was the same Health Department that was making an absolute nightmare of the Love Canal chemical-waste situation, a botched job of monumental proportions, and now they were going to help run a NIEMR-effects research program. The Health Department had been completely silent on the issue all the time the hearings were conducted in New York, but now, by gubernatorial fiat, they were made not only partners but the lead agency in the program.

It took eighteen months for the troika just to agree on who was going to be on the ten-person ARC panel. They split the panel among themselves like a kid dividing up his marbles: one for me, one for you, one for him.... The influence of PASNY and the Health Department prevented the naming of a truly outstanding panel. Becker, for instance, was rejected, even though he was better qualified than anyone else. The Chairman of the ARC panel became Michael Shelanski, a professor of pharmacology at New York University who had, in my opinion, no qualifications whatever for the job; he was a friend of David Carpenter's, the man at the Health Department who had been put in charge of organizing and running ARC.

In what must have been one of their first decisions, ARC decided to exclude me from even consideration for funding. They told me, in effect, that they wouldn't even read my proposal. ARC did approve about a dozen research proposals and, for the most part, they seemed to be well designed and above all honest. Unfortunately for the New York authorities, most of the studies were not relevant to the question of powerline NIEMR health risks because they were aimed at studying cells in test tubes rather than live ani-

97

mals—or humans. Thus, science in general would benefit, but I thought the people who were expecting concrete answers would be severely disappointed.

*

Well before the New York research program began, information began to trickle out about the studies that Dick Phillips was doing for Battelle, funded in 1976 by Robert Flugum of the Department of Energy (DoE). These studies, which according to Phillips were to "obtain a scientifically sound data base for establishing reliable and valid human exposure limits in order to ensure public safety and health," had been funded at a level exceeding that of all the studies ever done previously by all agencies put together.

Phillips's studies were flawed from the outset. For one thing, he chose to expose very large—that is, very *old*—rats. Thus his very choice of animal tended to rule out clear effects. Then he decided to house them in cages only four inches high, which squashed them into a volume much smaller than that mandated by Federal animal-care regulations. Finally, Phillips decided to use cages with a metal-mesh floor, which was supposed to eliminate what he believed to be a source of error in my studies. But because of the size of the animals and the cages, this meant that the male rats' testes were constantly in contact with a flow of electricity through the floor. (It was hardly surprising when his team later reported effects on the testicular hormones of the male rats.) Moreover, the wire-grid floors were totally inconsistent with the natural nest-building instincts of the female rat, which led to a long (and eventually unsuccessful) search for a litter material that would permit the animal some semblance of normal birthing conditions. Finally, he did his experiments at NIEMR levels ten times higher than those present in the real world. This tactic gave DoE and EPRI deniability: if by some chance the NIEMR *did* produce adverse effects in the rats, it could be argued that the levels were too high to be relevant to humans. So what Phillips had devised for the rat experiments was bound to fail.

Around 1977 Phillips and his investigators began traveling to various scientific meetings, delivering papers that basically reported no effects due to powerline NIEMR. But what Phillips and his people spoke and wrote about was not the data themselves but their *interpretation* of the data. Phillips alone had control over the actual data—he kept them locked in a safe—and it was he who made judgments about whether or not they showed effects. At the meetings, what was presented was always averages and summaries, never the data themselves.

I wanted the actual data, but when I wrote to Phillips for them, he said he would only sell them to me for copying costs: $20,000. That made me furious, so I wrote Flugum to demand the data, on the basis that I was qualified to examine and analyze them. The upshot of my demand was a classic Catch-22: as the law requires that a Federal agency turn over only what it has in its possession, and as Flugum had not received Phillips's data, and as he was going to rest content with Phillips's summaries and thus would *never* have the data, the government couldn't give them to me even though I had asked for them under the Freedom of Information Act.

But I did get copies of Phillips's monthly reports—after I had appealed

Flugum's decision all the way to Vice President Mondale. They told an interesting story.

In his December 1977 monthly report Phillips said, "We also plan to start the Marino three-generation study in January. It will take about 9 months to complete." Now obviously I had a big interest in this experiment; my name was on it—I had been the first American investigator to do such an experiment. As I read the monthly reports in the next year, it became clear that something was going wrong with Phillips's study, despite the fact that he had set it up using the best possible equipment and was running a simultaneous *replication* of the same study in another part of the Battelle lab.

In May Phillips reported that one generation of mice in one of the experiments was smaller than the other, but that because the experiment was being run "double blind" (i.e., in such a way that the technicians couldn't tell which group of mice was being exposed), he wouldn't know until December which group was smaller. In June he reported that the second-generation mice were now two weeks old, and in August he said that the third generation "will be born during the first week of September." But by the end of November, when the data were supposed to be in, things had suddenly changed. Phillips wrote, mysteriously, that "the system for exposing mice will be used continuously until the middle of March when the fourth generation of exposed mice will be six weeks old."

Now it was apparent to me that something unforeseen had happened, and that Phillips had changed the plan to include a fourth generation in the hope that whatever had happened would resolve itself. But apparently it didn't; in March Phillips reported that the *three*-generation experiment was complete and that the data seemed to show no effects. In April the experiment was ended altogether.

Almost a year later, in a report intended for government officials, Phillips reported for the first time *some* of the data from the "Marino experiment." His text, referring to a complicated table of average animal weights, suggested no effects from the NIEMR. But if one penetrated the table and statistically analyzed the data, what it showed was the following: the third-generation animals in one experiment were *heavier* than the controls, and those in the other were *smaller* than the controls.

There were only two possible conclusions from Phillips's data: (1) the experiments showed that NIEMR affects growth rate, and the apparent discrepancy between increased and decreased growth depended on factors that weren't controlled in the study; or (2) Phillips and his people had simply botched the job, despite all the money. But Phillips reached neither of these conclusions. What he did was to *average* the weights of the larger and the smaller mice and conclude there were no effects!

There was one further revelation at a 1982 meeting of DoE contractors. Phillips reported that a mistake had been made and that many experiments, including some in which he had reported NIEMR effects, had actually been done at a very low field strength—*120 volts/meter rather than 100,000 volts/meter*. This was a mistake roughly akin to that of the absentminded professor who went to class but forgot to put his pants on first.

*

99

In 1976 it had seemed we were in for big trouble down the road with grants. Then in 1978 I got a grant from NIH to do powerline research, and I began to think things were returning to normal. It is a big country and I would survive, even though there were people who disliked what I'd done. But in 1979 the screws began to tighten.

Part of it, I think, had to do with a *Saturday Review* story. A writer for the magazine who had done a previous piece on Becker's regeneration work called and said she wanted to do a piece on the New York hearings and their relationship to the NAS Sanguine committee. The result was "The Invisible Threat: The Stifled Story of Electric Waves," which appeared in September 1979. It went into detail about Matias and Sheppard, and about our charge that the NAS committee had been rigged. The article was very hard-hitting.

Philip Handler was outraged, livid. He called the author on October 1 and told her, "I am going to use every penny that we have in the National Academy of Sciences to break you and break the *Saturday Review*." That really upset her, and she called me to ask if I thought he could really do that. I told her not to worry because she hadn't written anything maliciously and that, besides, truth was a perfect defense to a defamation action.

But I wondered about myself. Maybe she didn't have anything to worry about, but I figured that I did. Philip Handler was the most powerful figure in the U.S. science establishment, and it was my ass he wanted, not *Saturday Review*'s. He took the position that the editor of the magazine had been duped by the writer, and the writer had been duped by me. Handler wanted the writer to hold a sort of press conference and confess that I had deceived her. She didn't, and *Saturday Review* refused to print a retraction. By the end of the year, things had simmered down, I thought.

But then in late January 1980 Handler sent a 17-page article to the editor of *Saturday Review* demanding that it be published, or else. His cover letter said: "We do, of course, have alternate means available to deal with the article.... As you may know, this institution has in the past successfully sought legal redress when one of our committees was slandered in the public prints and we would not hesitate to so commit our resources again, should that be necessary." He went on to say that the publication of the "enclosed article" would satisfy him, however.

In the article—which contained the typical language of Michaelson and Miller—Handler attacked the NIEMR investigators on whom I had based my views, and thus on whom the *Saturday Review* article had based its indictment of powerline NIEMR as a health risk. But his most florid language was reserved for me. He said my studies had been accepted by the writer "deliberately in disregard of the fact that they have been rejected as valueless by the rules by which science guards against shoddy work.... Are [Marino's experiments] believable? ... independent analysis of Marino's own data shows that there were no statistically *significant differences in the weight of the treated versus the untreated rats!* ... and that picture of the woefully stunted mouse? Perhaps the growth of some mice was indeed stunted, but it must have been a very small fraction of the total. And the experimental procedures ... most surely do not provide scientifically acceptable evidence that [NIEMR] causes such effects.

"Yet on this trivial dubious ground, the article in the *Saturday Review* built a case for a conspiracy in which are united the National Academy of Sciences, ... the federal government, the legal system, and for that matter any scientist who dares to disagree with Marino's claims. QED!"

Saturday Review did not publish Handler's article, nor did Handler sue. But to have someone as powerful as that so mad at me was not good. The point was driven home vividly when, in 1990, I saw Handler for the first and only time. I was in San Francisco to deliver an invited paper at the annual meeting of the American Association for the Advancement of Science. I passed a sumptuously appointed banquet room in the hotel, where fine ladies and gentlemen in formal attire were dining "by invitation only." I stood in the doorway and saw Handler standing there, talking with Philip Abelson, the editor of *Science*. They looked at me—hard—and all I could think about was how much better it had been when my antagonists had been simple men like Harvey and Matias.

*

In 1979 the laboratory in Syracuse really began to come apart. The process began when we sent three separate proposals to the VA for research support.

Unfortunately for us, by 1979 Marguerite Hays had risen to such a position within the Veterans Administration Research Service that *no* research proposal was outside her jurisdiction; this time there was no way around her. In the VA system, she had what was essentially the last word on each proposal. She gave our first proposal what was termed a "special" review and rejected it, apparently on her *personal* finding that it lacked scientific merit. The second proposal—one that I had written myself—was handled differently. She sent the proposal out to be reviewed by a researcher at Purdue University who was widely known by workers in our field to despise Becker personally; the two had clashed over questions of priority and they each had very strong feelings about the quality of the other's research on limb regeneration. It was therefore no surprise when the Purdue researcher wrote an excoriating review, which said that the proposal should be rejected because *anything* that was done in Becker's laboratory was worthless. On this basis, Hays turned down our second proposal.

The third proposal—the main one written by Becker himself—was handled still differently. As an established investigator, Becker was told that his application for continued support could be a "Part 3" application; that is, that it should describe the accomplishments of the past years, and describe in general terms the plans for future research. A Part 3 application frees the investigator from the necessity of locking himself into a specific research protocol and permits flexibility so that new and promising leads can be followed up; it is a privilege reserved for someone of known and established productivity. So Becker submitted a Part 3 application, and its review was personally handled by Dr. Hays. But in direct contradiction to her earlier instructions, she had the proposal reviewed as if it were a "Part 1" proposal—one calling for a great deal of experimental detail. Becker's proposal was rejected in October 1979 because it was "incompletely detailed."

Becker is a proud and private person and he did not, or could not, talk very much about what was happening to him. But it took its toll—on his health, on his general outlook about things, and on his will to continue fighting. He had truly pioneered in the use of electricity in medicine and biology, and like any other pioneer he'd had a rocky road. The Purdue researcher was only one of several who openly loathed him, and I think there were still others who felt that way but simply were more discreet about it. Research was doubtless Becker's first love, and he had stayed with the Veterans Administration all those years—not a prestigious job for an orthopedic surgeon—because one of the perks was the opportunity to do research.

In the past Becker had always seemed to find a way around the obstacles he encountered, and I guess it never dawned on any of us who worked with him that this resiliency was a function of age and the number of such obstacles. When he became eligible for early retirement in 1979 he began to think about it a lot. But he had spent twenty years building the laboratory and he wanted it to continue; so he tried to get a reading from Hays whether she would allow the laboratory to continue if he withdrew. In late 1979 she signaled her willingness to let us continue if Becker stepped aside. Courageously, he did—he retired from the Veterans Administration in June 1980, despite the fact that he was only 57 years old and at the peak of his scientific career.

Those of us who remained in the laboratory prepared a new proposal according to Dr. Hays's specific instructions. Becker was not listed on the proposal as an Investigator, but since the laboratory was supposed to be allowed to continue, Becker wanted permission to work in it—not *direct it*, merely *work in it*—as a volunteer to perform a single experiment involving limb regeneration. The experiment he wanted to do was the very same experiment which the head of the Veterans Administration, Max Cleland, had recently praised in a letter to the science reporter for the *Washington Post*.

At Christmas time of 1980 we were notified by Dr. Hays that the proposal, for the most part, had not even been reviewed, but that she had rejected it completely. In what in my experience is the all-time winner for hypocrisy, Hays said that one reason she was rejecting the proposal was that Becker had retired and thus the lab had lost its leader.

After being in existence for twenty years, the laboratory was ordered to close on eleven days' notice. I was told that my services were no longer needed in research, and was offered a job as an assistant hospital director. Within weeks after Hays closed our laboratory, she left the VA in Washington and took a job with a VA Hospital in Martinez, California.

Over the months and years that we had to deal with Hays, I must have asked Becker a hundred times about why he thought she was doing it to us. I thought that he might have insulted her or done something to her … something that would explain the vendetta she waged. He said there was nothing whatever in their previous relationship that he thought could conceivably justify what she was doing. I went through all the letters and correspondence between our laboratory and the VA in Washington—and anybody else that I thought might be involved—and I could not find a whisper of a reason for her antagonism. We discussed the possibility that she was Handler's hatchet, or the Navy's, but unless she is forced to respond under oath to questions about her relationship with Handler and/or the Navy, I'm afraid we'll never know.

*

102

It was a thrill to work in that lab all those years, to be on the cutting edge of some of the most important medical research in the country, and to have as a boss a man like Robert Becker. The plain fact is that we were done in by powerful people who could not stand to have us speak what we saw as the truth about the NIEMR health risk. As a result a great researcher was lost to science, and science itself was besmirched.

Whether what we accomplished in the PSC hearing was worth this loss, only history can tell.

CHAPTER 17

Confirmation: 1980–1985

In many respects, I grew up in Syracuse, New York. There I learned to be a scientist and a lawyer. There I met my wife, had four children, and bought my first home. There I gained experience and insight and learned lessons about life that are indelibly branded on me. The closing of the Syracuse laboratory was for these reasons a sad occasion. Near the end, I asked Becker if he had it all to do over again, what would he do differently? Only one thing, he said. He told me about an offer that he had received, ten years earlier, to move the entire laboratory to a university in Arizona. "I should have taken it," he said. "Everyone should move at least once in his career." Well, I had no choice; I was moving because the laboratory was closing. But it has turned out to be one of the best things that has ever happened to me.

I went looking for a job where I could do bioelectrical research, and found it as an assistant professor in the Department of Orthopaedic Surgery at Louisiana State University Medical Center in Shreveport. The attitude and environment in Louisiana and at LSU are as different from those in Syracuse as day from night. I am now regarded as an established scientist, whereas in Syracuse few saw me as anything more than what I was when I began, a graduate student. I had many enemies in Syracuse, most of whom I had never actually met or had anything to do with. They were Becker's enemies, and since I worked for him I was tarred with the same brush. (If I have any enemies at LSU, I earned them myself.) In New York, I was kept out of the mainstream of the hospital and its associated medical school. Even Becker had only limited interaction with the medical students, residents, and the various committees; those who worked for him had none whatever. During the seventeen years I was there, I worked with only one other faculty member (an anatomy professor) in a research project. At LSU, however, I became a full-fledged faculty member with responsibilities for teaching residents and medical students, and for participation in various medical school committees, as well as responsibilities for doing research, and in less than three years I've become involved in joint research or teaching projects with faculty members from almost every department in the school.

We bought a large antebellum plantation house and five acres of river-bottom land north of Shreveport in Belcher, Louisiana. I have a lifetime job in restoring it to its former grandeur, but I have a happy family to help me, and I figure I have exactly long enough to do it. Belcher is cotton country; many families there go back almost a century. Newcomers to Belcher are rare, and a newcomer who happens to be a Yankee, a scientist, a lawyer, an Italian, and a Catholic tends to stand out. There are no finer people anywhere than the people where I live, and their acceptance of me and my family, despite the obvious differences in our backgrounds, has been crucial to our happiness.

In January 1985 I was promoted to Associate Professor and given tenure at LSU. For the first time in my life I am at peace with myself, doing exactly the kind of work I want to do in a job that is beyond Federal political pressure, and living and working among good friends. I have to thank Sheppard, Matias, Handler, and Hays for this—a reversal I would never have believed could happen.

*

Just about the time Becker retired, bioelectricity entered a period of explosive growth. For numerous reasons—most of them obvious—the part of bioelectricity that deals with health risks of environmental NIEMR is unlikely to expand rapidly, but bioelectrical therapy is another matter.

In the beginning, the use of NIEMR for treating and diagnosing diseases was fueled principally with money from NIH and NSF, which are Federal agencies generally beyond even the appearance of untoward influence. (But even that is not always the case. In the late 1970s I was the only investigator in the United States who had Federal research dollars to study potential health risks of 60-Hz fields. My initial NIH grant led to many publications, but when I submitted a renewal application I was turned down. One member of the Study Section that voted down my renewal was a member of Dick Phillips's research team, and in fact was the primary investigator in Phillips's repeat of "Marino's three-generation study.") When the Federal seed money led to the confirmation of therapeutic effects from NIEMR, companies that sensed a viable product—and thus a profit—moved into the area. This process led to a variety of commercial devices, approved by the FDA in the late 1970s, for treatment of bone diseases. Some products were totally implantable, some involved the insertion of metal electrodes through the skin, and some had no actual physical contact at all with tissue but rather depended on pulsing magnetic fields to produce their effect. Today, NIEMR-producing devices are being evaluated clinically and in animal research with regard to their potential use for treating many more diseases than those of bone.

Our efforts at LSU are typical of the range of therapeutic applications of NIEMR now underway in various laboratories around the world. My colleagues and I are studying the use of NIEMR for accelerating normal fracture healing; measuring the natural electrical signals from the breasts of women with tactile lumps with a view to determining whether such a simple measurement can predict the existence of a cancer; and treating cancer in animals with electric currents, in an attempt to develop a technique for destroying localized cancerous lesions, a technique we hope will be ultimately developed for human use as well. We are conducting a clinical study to evaluate the use of an electrical technique for the treatment of bone infections, and we are planning a study involving the use of electroacupuncture for the treatment of lower back pain. We are studying the effects of magnetic fields on mammalian cells in tissue culture, hoping to unlock the actual mecha-

105

nism by which the cells are influenced by the NIEMR. And we are studying the effect of an electrical technique for accelerating healing of soft tissue such as skin, ligaments, and tendons. This work, and comparable work being done by many investigators in other laboratories, is intended to bring about a therapeutic result. If it works, or looks like it will work, funding can usually be obtained.

My *Handbook of Bioelectricity*, now in the last stages of publication by Marcel Dekker, will contain more than twenty chapters written by experts in various aspects of bioelectricity. It will be the first time that all the aspects of the field will have been contained within the covers of one book. And because of the incredibly rapid growth of bioelectricity, I expect it will be the last time that such a compilation can be made. I believe that bioelectricity will develop into a full-fledged science—the science of the 21st century.

*

The gist of the message that Becker and I delivered in New York—that NIEMR matters biologically—has begun to flower fully, then, in the area of therapeutic applications. In the area in which we actually verbalized our warning, that NIEMR exposure in the environment would make people sick, there have been virtually no "official"—by which I mean Federally funded and approved—studies regarding health risks. But resourceful individuals have managed to conduct such studies anyway and the results have turned out as predicted.

One day back in law school, a professor emphasized to us that "if the suspect walks like a drunk, talks like a drunk, smells like a drunk, and acts like a drunk, then the suspect *is* drunk." In the New York hearing, I pointed to more than fifty studies that described virtually every kind of adverse effect imaginable in animals from NIEMR exposure. Subsequently, when I reviewed the literature even more completely, I found an additional fifty studies of the same type and nature. (See R. O. Becker and A. A. Marino, *Electromagnetism and Life*, State University of New York Press, 1982.) It doesn't take a very bright person to recognize that people will react to NIEMR in essentially the same fashion as animals, and thus that people will be adversely affected by NIEMR exposure in the environment. Throughout the 1970s, essentially no studies had been performed on humans, and thus the extremist argument that there was no actual evidence that exposed individuals had been made sick by environmental NIEMR was, technically speaking, correct. But this is now changed. The human evidence is steadily mounting.

In the late 1970s Stephen Perry, an English physician, notified us that he had begun to see numerous cases of clinical depression among people who lived near overhead powerlines. We joined forces with Perry for a survey of several hundred locations of suicides, since suicide would be the most obvious result of severe depression. We actually measured the NIEMR (the magnetic component specifically) at the locations of suicides and found that the areas with the highest NIEMR levels showed many more suicides than areas chosen randomly. These results were reported in *Health Physics* in 1981.

One of the most disturbing reports was made in 1979 in the *American Journal of Epidemiology*. Nancy Wertheimer and Ed Leeper, of the University of Colorado Medical Center in Denver, found in a survey of several hundred homes that the death rate from cancer for children under 19 living in homes close to high-current power distribution lines was twice what was expected. In 1982 they reported (in the *International Journal of Epidemiology*) a similarly significant correlation between powerlines and adult cancer.

The same year, W. E. Morton reported excess cancer risk among housewives in Oregon who lived in houses with radiant electric heating, which produces NIEMR. Also in 1982 a Swedish group headed by L. Tomenius found a correlation between cancer in juveniles and proximity to high-voltage powerlines in the Stockholm area. In 1984 a strange cluster of rare ovarian tumors was found in five young girls living near a 69,000-volt line in Florida; at the time of the report, four of the girls had died.

In 1982 John R. Lester and Dennis F. Moore, writing in the *Journal of Bioelectricity*, described an apparent association between cancer and NIEMR in Wichita, Kansas. They found that the incidence of cancer tended to be more frequent along terrain crests, which would receive the highest levels of NIEMR, as compared to cancer levels in the valleys or low-lying areas. Cancer incidence appeared to be related to the probability of exposure to NIEMR from radar. In a second study, also published in 1982 (reconfirmed in 1985 after the conclusion was challenged by the U.S. Air Force), the same authors examined the incidence of cancer in the U.S. counties containing the cities nearest each of the ninety-one Air Force bases. They found that both men and women living in the test counties were more likely to get cancer than were people living in similar counties not located near Air Force bases. Thus they showed an association between some factor emanating from Air Force bases—their hypothesis was NIEMR—and the incidence of cancer. Finally, in May 1984 Wertheimer reported at a meeting of the American Association for the Advancement of Science in New York City her preliminary findings that showed a relationship between electric blanket use and fetal loss.

In addition to these residentially located effects, reports have begun to multiply rapidly in the last five years on effects among workers in electrical occupations. Samuel Milham Jr., the director for occupational health and safety in Washington state, reported in the *New England Journal of Medicine* in 1982 that workers in power stations, aluminum smelters, and other high-strength NIEMR environments suffered an excess risk of cancer, particularly acute leukemia (the same type Wertheimer and Leeper found in children). Also in 1982 W. E. Wright and his co-workers reported in the British medical journal *Lancet* that there was a similar increased risk for electrical workers in the Los Angeles area. M. E. McDowell found such correlations in England and Wales in 1983, and so did M. Coleman in the same year. Both reports were published in *Lancet*. Two of the latest reports (1984) have shown connections between brain tumors and NIEMR exposure in workers in Maryland, and an increase in eye cancer among male electrical and electronics workers in England and Wales.

Similar connections are being found between cancer and higher-frequency NIEMR of the kind that is emitted by microwave devices and radar. This

107

year the most extensive study ever done on humans, among military personnel in Poland who are exposed to NIEMR, has found a *tripling* of the expected incidence of cancer, with the risk highest for blood-forming organs,
the lymphatic tissues, and the thyroid gland. This study, which analyzed all
cancer cases in the Polish military reported between 1971 and 1980, was
conducted by Dr. Stanislaw Szmigielski of the Center for Radiobiology and
Radioprotection in Warsaw, and confirmed earlier work he did that found
cancer in animals from NIEMR exposure. It also confirms a recent American lab study, funded by the Air Force, which apparently found high cancer
incidence in rats exposed over a long period to high-frequency NIEMR. In
addition to these results, numerous clusters of significant miscarriage rates
have been reported among women who work at video display terminals in
the USA and Canada. (In response to these reports, a large study involving
over 10,000 women is being undertaken by Dr. Irving Selikoff and associ-
ates at Mt. Sinai School of Medicine in New York.)

Health problems have also been reported in people who live near PAS-
NY's 765,000-volt powerline. The various reports are anecdotal in nature,
but they have been persistent and serious enough to have warranted a series
of six long and detailed news stories in the Rome (N.Y.) *Daily Sentinel*,
written by John Golden. Beginning around 1981, about two years after the
line was energized, four families who lived relatively close to the line began
to report nonspecific complaints such as headaches, listlessness, and vision
trouble, as well as difficulties with their animals. (Similar complaints of
livestock problems came from other areas near the line as well.) As time
went on the problems worsened. One family which kept accurate records of
their livestock reported chicken eggs not hatching, birth and milk production difficulties with cattle, and bizarre deaths of several types of animals,
until finally their farm operations went precipitously downhill. Then they
reported the development of a rare form of thyroid disease in the daughter,
and soon after, the mother developed a thyroid condition as well. In another
family the father, who had cancer when the line went into operation, was
reported to have declined swiftly after that; he died in 1982. In a third family a daughter's heart condition worsened, and she had to undergo surgery.
All the early nonspecific complaints have persisted as well, according to
Golden's reports. One family took PASNY to court and obtained a settlement of $20,000, after PASNY had reportedly spent $200,000 in its defense.

*

What has the industry and government done to deal with the NIEMR
problem?

The chief industry response was to fund the work of Dick Phillips at
Battelle. In 1976 he and his chief engineer visited me in my laboratory in
Syracuse, and we talked about the work he was planning. He was courtly
and polite, but he sat there with a big Cheshire cat grin, and I felt very much
like Tweety Bird. I had a few dollars for research, he had power-company
and Department of Energy millions. I had no influence, but he had (or shortly would have) connections with the National Institute of Environmental
Health Sciences, the Environmental Protection Agency, and the Navy. (In

1985 he went to work for EPA at its Health Effects Research Laboratory at Research Triangle Park, N.C.) He was going to duplicate my studies, show up my flaws, publish the results, and squash me like an ant. But that's not the way it turned out. In the subsequent nine years he spent tens of millions of dollars (only he knows for sure how much), but his experiments were essentially one boondoggle after another and he accomplished very little. In 1984 I went to the annual meeting of the Bioelectromagnetics Society in Atlanta and delivered what I hope will be the epitaph to his studies. I told him, and the audience, point by point why his studies were a failure (see Chapter 16).

The program in New York was the only state-level research program dealing with NIEMR, but at the outset it was skillfully subverted by PAS-NY. One of the tactics that I perceived in research support in the NIEMR area, going all the way back to the 1940s, was the idea of funding experiments that could foster the goals of the funder, but that lacked the potential to hurt the funder. In the Tri-Service program in the late 1950s, for example, no studies were done at NIEMR levels that were commonly present in the environment, so the results could not show that actual exposure patterns were unsafe. In the ensuing years, most Federal and industry research funds were spent for studies involving mathematical modeling, use of unrealistically high exposure levels, and other kinds of studies that could not possibly be related to actual, real-life exposure conditions. Unfortunately, the New York ARC has done the same thing. They have commissioned studies that mostly deal with cells in test tubes. It simply doesn't matter what the cells do: they could turn from malignant to benign or vice versa, they could die, swell to ten times their size, or sit up and play Dixie. The argument will have been preserved that these are *only cells*, that such work is not relevant to human exposure, and therefore that the data generated do not apply to the setting of human safety standards.

The ARC was supposed to generate a body of data that administrators and legislators in New York, and other states, could use to protect the public health. The public doesn't care about mechanisms, or scientific mumbo-jumbo, or about how many cells can sit on the head of a pin. They want to know if living within 1000 feet of a powerline is going to make them sick. The New York program was supposed to prove or disprove this conclusion, but it is scheduled to go out of business sometime in 1986, and it will do so with a muted whimper. It will conclude that maybe there is a problem, maybe there isn't, they can't be sure, but if they could have $5 million more and five more years, maybe they would be able to give an answer.

At the Federal level the situation remains an electric wilderness. A loose organization of what might be called the lead Federal agencies has been formed, consisting of the Department of Defense (particularly the Navy), the Environmental Protection Agency, the Department of Energy, and the National Institute for Environmental Health Sciences. These are the folks with the money. They don't deny the existence of a problem with NIEMR; they have put considerable distance between themselves, collectively, and the Michaelson/Schwan view. This represents an enormous change from "official" thinking in the 1960s and 1970s, and is obviously a step in the right direction. The problem with the people in this self-organized group (many of whom I respect as individuals) is that they are trying to do the impossible—*make* scientific results conform to the shifting priorities and perceptions of their individual agencies.

Although the interactions of individuals and agencies at the Federal level remain too complex today to be described here, the most important aspect of Federal activity is relatively clear. The Federal government has accomplished virtually nothing with regard to NIEMR health effects since the problem matured after World War II. After all these years and all the millions, they have not proved or disproved the existence of a NIEMR risk, expanded our knowledge of how NIEMR affects living systems, narrowed the focus of inquiry, sharpened our understanding or, so far as I can tell, accomplished any other significantly useful purpose.

What we predicted in the 1970s regarding health risks of environmental NIEMR has come to pass. Today, if you ask a power-company representative if there is a problem, he will of course deny it. If you ask a Federal official the same question, you can expect an evasive reply. If you ask someone who doesn't owe his job or research support to either group, you will probably be told that there is in fact a problem—which is exactly what Becker and I said in 1974.

I think we need a serious, sustained Congressional investigation. There are a handful of men and women who must be called to account for their stewardship of the nation's civilian and military health. In the absence of a Congressional housecleaning and a Congressionally mandated design for a goal-oriented program of NIEMR research, the NIEMR health problem will simply sink deeper into the 1918-type trench warfare that exists at present between Federal officials and independent scientists.

Selected Reading List on Biological Effects of Electromagnetic Energy

Books

Stanislaw Baranski and Przemyslaw Czerski, Eds., *Biological Effects of Microwaves*, Stroudsburg, Pa.: Dowden, Hutchinson & Ross, 1976.

R. O. Becker, Ed., *Mechanisms of Growth Control*, Springfield, Ill.: C. C. Thomas, 1981.

R. O. Becker and A. A. Marino, *Electromagnetism and Life*, Albany, N.Y.: SUNY Press, 1982.

R. O. Becker and Gary Selden, *The Body Electric: Electromagnetism and the Foundation of Life*, New York: Morrow, 1985.

P. M. Boffey, *America's Brain Bank: An Inquiry into the Politics of Science*, New York: McGraw-Hill, 1975.

Paul Brodeur, *The Zapping of America: Microwaves, Their Deadly Risk and the Cover-up*, New York: Norton, 1977.

H. S. Burr, *The Fields of Life*, New York: Ballantine, 1972.

B. M. Caspar and P. D. Wellstone, *Powerline: The First Battle in America's Energy War*, Amherst, Mass.: University of Massachusetts Press, 1981.

Jane Clemmensen, *Nonionizing Radiation: A Case for Federal Standards?* San Francisco: San Francisco Press, 1984.

A. P. Dubrov, *The Geomagnetic Field and Life*, New York: Plenum Press, 1978.

L. I.. Klessig and V. L. Strite, *The ELF Odyssey: National Security Versus Environmental Protection*, Boulder, Colo.: Westview Press, 1980.

Herbert König, A. P. Kreuger, Siegnot Lang, and Walter Sonning, *Biologic Effects of Environmental Electromagnetism*, New York: Springer, 1981.

Boguslaw Lipinski, Ed., *Electronic Conduction and Mechanoelectrical Transduction in Biological Materials*, New York: Marcel Dekker, 1982.

J. G. Llaurado, A. Sances Jr., and J. H. Battocletti, *Biologic and Clinical Effects of Low-frequency Magnetic and Electric Fields*, Springfield. Ill.: C. C. Thomas, 1974.

E. J. Lund, *Bioelectric Fields and Growth*, Austin, Tex.: University of Texas Press, 1947.

Karel Marha, Jan Musil, and Hana Tuha, *Electromagnetic Fields and the Life Environment*, San Francisco: San Francisco Press, 1970.

Allan Mazur, *The Dynamics of Technical Controversy*, Washington, D.C.: Communications Press, 1981.

New York Academy of Sciences, *Electrically Mediated Growth Mechanisms in Living Systems*, New York: N.Y. Academy of Sciences, 1974.

M. A. Persinger, Ed., *ELF and VLF Electromagnetic Field Effects*, New York: Plenum Press, 1974.

A. S. Presman, Electromagnetic Fields and Life, New York: Plenum Press, 1970.

Margaret Rowbottom and Charles Susskind, *Electricity and Medicine: History of Their Interaction*, San Francisco: San Francisco Press, 1984.

Hans Selye, *Stress*, Montreal: Acta, 1950.

N. H. Steneck, *The Microwave Debate*, Cambridge, Mass.: MIT Press, 1984.

N. H. Steneck, Ed., *Risk/Benefit Analysis: The Microwave Case*, San Francisco: San Francisco Press, 1982.

F. G. Sulman, *The Effect of Air Ionization, Electric Fields, Atmospherics, and Other Electric Phenomena on Man and Animal*, Springfield, Ill.: C.C. Thomas, 1980.

Albert Szent-Gyorgyi, *Introduction to a Submolecular Biology*, New York: Academic Press, 1960.

Albert Szent-Gyorgyi, *The Living State*, New York: Academic Press, 1972.

M. S. Tolgskaya and Z. V. Gordon, *Pathological Effects of Radio Waves*, New York: Consultants Bureau, 1973.

U.S. Department of Energy, *Biological Effects of Extremely Low Frequency Electromagnetic Fields* (Proc. 18th Ann. Hanford Life Sciences Symp., Richland, Wash.), Washington, D.C.: DOE Technical Information Center, 1979.

L. B. Young, *Power Over People*, New York: Oxford University Press, 1973.

Articles

R. O. Becker, Brain pollution, *Psychology Today*, February 1979.

R. O. Becker, Electromagnetic forces and life processes, *Technology Review*, January 1972.

R. O. Becker and A. A. Marino, Electromagnetic pollution, *The Sciences*, January 1978.

P. M. Boffey, Project Seafarer: Critics attack National Academy's review group, *Science*, 18 June 1976.

John Golden, articles on PASNY powerline in Rome (N.Y.) *Daily Sentinel*, 4 April 1981, 3 March 1982, and 9, 16, 25, and 27 October 1982; update in *Watertown Daily Times*, 20 April 1984.

L. J. Hertzel, Anybody want a copy of my prayer? A matter of power, *North American Review*, Spring 1978.

E. J. Lerner, The drive to regulate electromagnetic fields, *IEEE Spectrum*, March 1984.

E. J. Lerner, Biological effects of electromagnetic fields, *IEEE Spectrum*, May 1984.

A. A. Marino and R. O. Becker, High voltage lines: Hazard at a distance, *Environment*, November 1978.

A. A. Marino, Electromagnetic fields and public health, *Assessment and Viewpoints on the Biological and Human Health Effects of Extremely Low Frequency (ELF) Electromagnetic Fields*, American Institute of Biological Sciences, Washington, D.C., May 1985.

Allan Mazur, A. A. Marino, and R. O. Becker, Separating factual disputes from value disputes in controversies over technology, *Technology in Society*, vol. 1, 1979.

Kathleen McAuliffe, I sing the body electric, *Omni*, November 1980.

Kathleen McAuliffe, The mind fields, *Omni*, February 1985.

Dennis Meredith, Healing with electricity, *Science Digest*, May 1981.

J. T. Mulder, Short-circuited by the Establishment, *Empire Magazine*, Syracuse (N.Y.) Herald-American, 15 February 1981.

C. G. Park and R. A. Helliwell, Magnetospheric effects of power line radiation, *Science*, 19 May 1978.

Lowell Ponte, The menace of electric smog, *Reader's Digest*, January 1980.

Joel Ray, The hazards of high wires, *The Nation*, 18 February 1978.

Joel Ray, Citizens protest high power lines, *The Bulletin of the Atomic Scientists*, April 1980.

Joel Ray, PASNY, PSC tangle over energy hazards, *Empire State Report*, 16-30 November 1980.

Susan Schiefelbein, The miracle of regeneration: Can human limbs grow back?

Saturday Review, 8 July 1978.

Susan Schiefelbein, The invisible threat: The stifled story of electric waves, *Saturday Review*, 15 October 1979.

N. H. Steneck, H. J. Cook, Arthur Vander, and G. L. Kane, The origins of U.S. safety standards for microwave radiation, *Science*, 13 June 1980.

Task Force of the Presidential Advisory Group on Anticipated Advances in Science and Technology, The science court experiment: An interim report, *Science*, 20 August 1976.

P. D. Wellstone and Lamont Tarbox, Confrontation on the prairie, *The Progressive*, December 1977.

Lally Weymouth, The electrical connection: Part I, *New York Magazine*, 24 November 1980.

Lally Weymouth, The electrical connection: Part II, *New York Magazine*, 1 December 1980.

L. B. Young, Danger: High voltage, *Environment*, May 1978.

L. B. Young and H. P. Young, Pollution by electrical transmission, *The Bulletin of the Atomic Scientists*, December 1974.

*

In addition to the above books and articles, the reader is encouraged to consult *Microwave News*, *The Journal of Bioelectricity*, and the *Newsletter* of the Bioelectromagnetics Society for continuing developments.

Name Index

Pollack, Capt. Charles, 70–72

Rather, Dan, 70–72
Rheingold, Arthur, 11–12, 36, 62, 79

Schwan, Herman, 13–17, 23–25, 27–28, 30–33, 36–38, 41–48, 54, 56, 60, 65, 81, 89, 94, 110
Selden, Gary, 3
Selikoff, Irving, 108
Semmelweis, Ignaz, 94
Shah, Kanu, 66
Shelanski, Michael, 97
Sheppard, Asher, 82–83, 95–96, 100, 105
Simonds, Betsey, 76
Simonds, Rev. Robert, 76
Simpson, Robert, 6–8, 9, 11–12, 29, 31, 35–46, 48–49, 52–53, 55, 62, 64–66, 73–74, 77–81, 87, 93, 97

Southern, William, 44
Spieler, Cliff, 68
Steneck, Nicholas, 88
Swidler, Joseph, *iv*
Szent-Gyorgyi, Albert, 2
Szmigielski, Stanislaw, 108

Tomenius, L., 107
Tunney, Sen. John, 17
Tyler, Cmdr. Paul, 1, 3, 6, 7

Wallace, Francis X., 35, 51–52, 55, 64, 73–74, 78–79, 85, 96
Wallace, Mike, 70, 72–73, 75
Wertheimer, Nancy, 107
Wever, Rutger, 43
Wright, W. E., 107

Zaret, Milton, 17

www.ingramcontent.com/pod-product-compliance
Lightning Source LLC
Chambersburg PA
CBHW032006190326
41520CB00007B/380